Rubber Science

Yuko Ikeda · Atsushi Kato
Shinzo Kohjiya · Yukio Nakajima

Rubber Science

A Modern Approach

 Springer

Yuko Ikeda
Center for Rubber Science and Technology,
    Faculty of Molecular Chemistry and
    Engineering
Kyoto Institute of Technology
Kyoto
Japan

Atsushi Kato
Department of Automotive Analysis
NISSAN ARC, LTD.
Yokosuka, Kanagawa
Japan

Shinzo Kohjiya
Kyoto University
Kyoto
Japan

Yukio Nakajima
Department of Mechanical Science and
    Engineering, School of Advanced
    Engineering
Kogakuin University
Hachioji, Tokyo
Japan

ISBN 978-981-10-9744-7   ISBN 978-981-10-2938-7   (eBook)
https://doi.org/10.1007/978-981-10-2938-7

Printed on acid-free paper

This Springer imprint is published by Springer Nature
The registered company is Springer Nature Singapore Pte Ltd.
The registered company address is: 152 Beach Road, #21-01/04 Gateway East, Singapore 189721, Singapore

Preface

Rubber is polymeric or macromolecular, and rubber science is unquestionably an important branch of polymer or macromolecular science. Note that the two words, macromolecule and polymer, are interchangeably used widely. In most polymer science textbooks, however, there has not been adequate description of rubber to appropriately introduce readers to rubber science. If any, brief mention of rubber research by H. Staudinger, the first Nobel Prize winner from the macromolecular science arena, is made in an introductory chapter owing to the crucial contribution of rubber research in establishing the macromolecular theory. His researches on rubber were decisive in having negated the colloidal association theory of macromolecules, together with his tireless argument that one covalently bonded macromolecule could be colloidal (without any association) due to its high molar mass, i.e., macromolecular theory.

One reason for such neglect of rubber science at present might be because rubber science has been regarded as a traditional scientific area highly oriented to technical applications. Two historical events, i.e., invention of rubber vulcanization by C. Goodyear in 1839 and the beginning of mass production of pneumatic rubber tires early in the twentieth century, occurred well before the establishment of polymer science in the 1940s. That is, the rubber industry was established without much systematic assistance from the achievement of polymer science. In this historical context, the progress in rubber science has been highly dependent upon huge number of trial-and-error trials, often not much based on any modern scientific methodology for research and development.

However, the maturing of relevant disciplines including polymer and analytical science has recently created a new trend among the traditional sciences including rubber. That is, the utilization of modern and more or less sophisticated techniques has stimulated the spread of state-of-the-art methods to be used in rubber research at both university and private company laboratories. Increasing numbers of scientists and technologists are now interested in rubber not as a traditional field but as a promising one for applying the most modern scientific achievements, both theoretically and experimentally. Unfortunately for them, however, few books describing such latest scientific achievements on rubber are available, because almost all textbooks on rubber tend to describe conventional and traditional results in detail not much on the modern approaches. At the same time, this shortcoming in essence may be a continuing negative aspect of rubber science that may continue well into the middle of the twenty-first century, against its necessary and possible contribution to the sustainable development of the modern transportation society in this century.

This volume is the first trial to overcome that deficiency. It is neither conclusive nor comprehensive, but it may serve as a pilot version to meet the recent demand particularly among individuals working in rubber. It includes a few structural topics that have been disseminated in this century using X-ray scattering techniques from a modern synchrotron facility and neutrons from a nuclear reactor, plus the most recent advanced studies in the mechanics of tire manufacture. The results elucidate both the network structure and the vulcanization mechanism. Further, the most recent three-dimensional imaging technique applied to transmission electron microscopy, i.e., 3D-TEM, is used to elucidate nanofiller distribution in the rubbery matrix, which may give rise to important results for revealing the mechanism of rubber reinforcement. These techniques have recently been fundamental in lots of scientific areas and are the most urgent ones in rubber arena. The authors have done excellent work for the systematic presentation of these recent achievements, not simply to mentioning them as technical examples.

We are confident that this pilot version of a modern resource on rubber is extremely worthwhile for the future progress of rubber science in the twenty-first century. We hope that our trial efforts here will be soon followed by similar ones but from a different standpoint than ours. We are hopeful, too, that ours and other such versions will, in combination, accelerate the progress of rubber science, ultimately resulting in its much greater contribution to the sustainable development of the transportation society during this century.

Kyoto, Japan   Yuko Ikeda
Yokosuka, Japan   Atsushi Kato
Kyoto, Japan   Shinzo Kohjiya
Hachioji, Japan   Yukio Nakajima
June 2017

Contents

Chapter 1
Introduction to Rubber Science

1.1 Rubber and Elastomer

1.1.1 Materials and Matters

Just before studying rubber science which is a branch of materials science, we have to know the difference between material and matter. Scopes of the two words are shown in Fig. 1.1 as per the set theory, which suggests that the matter forms a larger set (ensemble) than the material. This figure also indicates that any materials are contained in the matter. (In Fig. 1.1, fact and information are shown, too, which are to be explained later.) Then, leaving the mathematical set theory, we have to ask what is the indicator or the index to qualitatively differentiate the two words? The answer is the social utility for us. In other words, usefulness is an essential property of materials, while the matter may be either of use or of no use [1–3]. In our everyday life, we use thousands of products made from various materials. Materials are either naturally occurring or synthetic. The latter materials are usually manufactured industrially via mechanical and chemical conversion processes. Surely, one of the hallmarks of modern industrialized society after the Industrial Revolution is our increasing extravagance in using synthetic materials. Currently, materials science is regarded to be among the three major technologies together with information science and biotechnology. Or, more recently, major technologies would be four, when environmental science is to be included.

On the other hand, the essence of matter (or often "substance" is used in chemistry arena) is the objective existence itself in this world, in accord with a priori assumption that the majority of natural scientists, technologists, and engineers hold unconsciously or without much thinking [1, 4–7]. The existence of matter does not have the raison d'être. In other words, it does not need to have a justification for existence [5–7], which is one of the basic features of modern science. In addition, the matter is philosophically assumed to be indestructible as a whole [6, 7]. That is, matter may be converted to the other matter, but be never nullified. This

© Springer Nature Singapore Pte Ltd. 2018
Y. Ikeda et al., *Rubber Science*, https://doi.org/10.1007/978-981-10-2938-7_1

Fig. 1.1 Relationship
between material and matter

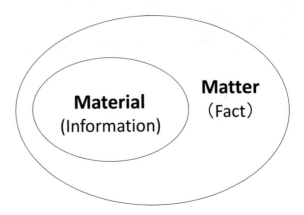

philosophical understanding is related with scientific establishment of the atomic theory and the law of energy conservation by the end of nineteenth century [4, 5, 7–10]. The law was extended to accommodate matter or substance as well as energy, through the Einstein's relationship:

$$E = mc^2$$

where E is energy, m is mass, and c is the velocity of light. By the modern scientific understanding, materials are matters of use for the human beings, as explained by Fig. 1.1. In this book, rubber is described as a matter from a basic scientific standpoint, and it is also treated as a material from materials science viewpoint.

1.1.2 Materials Have Afforded the Grouping of the History of Us

Since the materials are useful for us, history of the human beings has been divided into three periods by what kind of material is mainly used at each era: the Stone Age, the Bronze Age and the Iron Age. About seven million years ago, an ape-man (*Pithecanthropus*) appeared on the earth, and it evolved into a hominid, i.e., the primitive human race, about two million and a half years ago. The human race began using a flake tool made from the stones. That is, the Paleolithic Age was started, followed by the Neolithic Age when a polished stone tool became popular. During the Stone Age, the human race learned the use of fire at about two million years ago, not only to warm their bodies but also to cook various food materials later. The importance of this event gave rise to a Greek myth that Prometheus, who had stolen the heavenly fire and gave it to a man, was punished to be chained to the mount Kavkaz by the order of Zeus. He was liberated only by Zeus's son, Herakles. Cooking a meal is basically a process of chemical change by using the fire, with some physical or mechanical changes of food materials. Combined with the invention of earthen wares as a container, the cooking extended the range of food materials and the storage time of food. Through these discoveries and inventions, the hominids were evolved into *Homo sapiens*, the present human beings. And eighty to seventy thousand years before, they began to spread worldwide from the East African Valley [11–13]. The odyssey might have ended when they arrived at the southern tip of the South American continent around twenty thousand years before.

The last glacial era was ending about twelve thousand years before the Christ (BC), when culturing of grains, specifically wheat, was started at the Crescent in the Middle East. This is the beginning of a full-scale agriculture [12–14]. Progress of agriculture necessitated more and more sophisticated stone and earthen wares. The agriculture also made constant migration of people unnecessary, to result in settling down of people permanently at a suitable place for agriculture. When they had been chasing animals and collecting fruits and nuts, they were obliged to move constantly or at least seasonally. The grains were much preservable among food materials and stored in an earthen vessel. Therefore, the supply of food became much more stable under the agricultural economy than under the previous hunting and collecting economy. The progress of agriculture at the Neolithic Age gave way to the civilization. Thus, via the short Bronze Age, the Iron Age began about four thousands BC. The spread of ironwares was, particularly, effective in bringing about civilization among many human races [15–20], together with agricultural production of food materials.

The importance of ironwares in a society is due to the versatile nature of iron as a material, and its significance is understood by the fact that the society of not yet using ironwares is often called "uncivilized." Additionally, using ironwares has enabled mankind to make use of various materials other than iron. Under these diversification trends, the Mayan and Inca civilizations, where the ironwares were not so widely used, discovered natural rubber (NR) and manufactured rubber balls [21]. The games using the rubber ball might be of religious and/or political meanings. It is notable that Mayan and Inca civilizations were lacking ironwares and their diets lacked any wheat. Their main diets were maze and potato. Therefore, the civilizations were holding a unique position: The use of NR may be one of their peculiarities. The utilization of NR in these civilizations was, however, not much technical, and the use had been quite limited even after the encounter with European people in the sixteenth century. NR had to wait for the invention of vulcanization in 1839 by C. Goodyear (see Sect. 2.2) for its wide technical uses. In combination with the invention of pneumatic rubber tires, the rubber has been developed into an indispensable material supporting the modern transportation society, up to now [21].

The above discussion has been based on the general idea that rubber science is a branch of material science, and our history so far may be that of materials. A little mentioned transportation society is assumed to be the last stage of Materials Ages to continue to the end of this century [21], followed by the Information Age. In other words, "We are moving from the Age of Materials to that of Information" [22]. In this book, we still assume the former would be the case, at least to the end of this century. Tendency toward the Information Age might have already been recognized, coexisting with the Age of Materials. For a moment, however, materials are still more important: The most recent trend of Internet of Things (IoT) gives a support to the above recognition.

1.1.3 Science on Rubbery and Elastomeric Materials

In the twentieth century, demand of rubber for automobile tires was highlighted with the spread of automobiles: Famous Ford Model-T was marketed as early as in 2008. The price of NR was skyrocketed, which induced a strong interest in chemical preparation of synthetic rubbers, at synthetic chemistry arena [21]. Full-scale development of synthetic rubbers, however, was observed at the 1930s and the 1940s, and urgent, mass-scale developmental activities of synthetic rubbers were followed by their industrial mass production during the World War II, which did decisively open the era of present coexistence of NR and synthetic rubbers [21, 23–25]. With the increase of rubber use, the word "elastomer" has gained a frequent usage particularly in the synthetic rubber arena [24].

Quite often elastomer has been used interchangeably with rubber. However, elastomer, the origin of which is an adjective word "elastic," means elastic materials, and it includes cross-linked rubber and thermoplastic elastomer (TPE, see Sect. 3.5.3). On the other hand, rubber includes raw rubber, rubber adhesives, and rubber glues, which are not cross-linked rubber products. More importantly, rubber is used in the technical words expressing the state and property of matter, such as "rubber elasticity", "rubbery state", and "rubbery matrix". Hence, rubber is a much more versatile word than elastomer.

At present, science or learning, which is nowadays almost equivalent to science, is broken down as shown in Table 1.1.[1] (Some humanity and social scientists still do not like to use the word "science". They prefer "learning" or "Wissenschaften" in German. For them, science means natural science only). The following two points are noteworthy in the table:

(i) Philosophy is a science to study how to understand this world and the mankind in general, while mathematics is concerned with quantity of all that are extant in this world (including our brain or mind). Hence, the two are an independent part of science. This claim is nowadays more acceptable: Philosophy on science and technology is recognized as one of the most important areas in the modern philosophy, and the use of mathematics has been increasingly popular in lots of human and social science areas, too.

(ii) Engineering, agriculture, and medicine have been recognized among natural science so far. However, they are placed separately under technology in the table. They are on technics, but science on technics had not been considered to be independent; they had been considered as an application side of mathematics, physics, chemistry, or biology.

Table 1.1 Breakdown of science in accordance with object of the study

Philosophy	View of the world	Natural science	Nature
Mathematics	Quantity of everything	Physical science	
Humanity science	Humanity	Physics	
Historiography		Chemistry	
Geography		Astronomy	
Literature		Geoscience (geology)	
Psychology		Biological science	
Social science	Society	Biology (botany, zoology)	
Economics		**Technology**	Technics
Jurisprudence		Engineering	Industrial technique
Sociology		Agriculture	Agricultural technique
Pedagogy		Medicine	Medical technique

This is still a traditional tendency, which is surely due to the historical reason [26, 27]. In the classical society (Greek and Roman), technical works were handled by slaves, not at all by the free citizens. Then, the philosophers (roughly equivalent to scientists nowadays) tended to be proud of not engaging in technical works: The spiritual activity, i.e., metaphysics, is for the citizens, while physical labors are for only slaves, not for citizens. This prejudice had persisted through the medieval period, and general recognition of the importance of technics was delayed until the progress of the Industrial Revolution. Steam engine was invented to give rise to industrial manufacturing, which essentially demanded the presence and supervision of engineers, not simply a manager [4, 17–20]. At the final stage of Industrial Revolution, technology, a science on technics, was recognized to be a separate subject from the natural science arenas: Technology is the science concerned with technics that are mediating between nature and the mankind (see Fig. 1.3 shown later), and it should be based on human and social sciences as well as on natural science.

Here, it is notable that the social importance of technology was first recognized not in Britain where the Industrial Revolution was developed first, but in France, then in Germany and in the USA, followed by Japan. These countries were trying to catch up Britain in terms of industrialization, and in need of as many engineers as possible. In France, the revolutionary government established *Ecole centrale des travaux* publics in 1794 in order to educate military and civil engineers, which was renamed to be famous "*Ecole polytechnique*" in 1795. "*Technische Hochschules*" and A & M (agriculture and mechanical engineering) universities were established in Germany and in many states of the USA, respectively, during the 2nd half of the nineteenth century. In Japan, a few faculties of engineering and agriculture were set up in its Imperial Universities by 1900. In Britain, many traditional universities which were established long before the Industrial Revolution (including Cambridge and Oxford) did not like to include technological colleges, since technology was regarded as the technics itself: Technology was not for learning, but was supposedly the skills simply to be practiced, for the majority of professors then, particularly in England.

Recently, the requirement of independent technology is renewed, i.e., technology as an independent science has increasingly been acknowledged by many people due to the importance of sustainable development (SD) for the future of us. In other words, further development of technology is the most urgent assignment on us for the success of SD of our modern society. In Table 1.2, a few technologies that have been established or are establishing at present are shown. The rubber science described in this book is a branch of materials science (see Sect. 3.1 for the origin of materials science), where nanotechnology is recently in a fashion as a promising subject. Some impressive results at nanoscale are presented in this book on rubber, too.

Information, an object of information science, is shown in Fig. 1.1. Facts in the world are too many to account all. Among so many facts, we have been trying to find out or to select those of much use for us to get knowledge. Then, the assembly of such knowledge is the basics for us to think. "I think, therefore I am" is correct, and it is essential for our thinking to have an assembly of the knowledge of useful facts. Both information and material are assumed to be useful for the human beings, and in order to judge their utility for us and for society, it is needed to place their science at technology arena. This process of choosing matter and fact of use, i.e., selecting materials and information, necessitates good considerations based on social and humanity studies. This is the reason of our classification materials science (and information science) into technology, not into natural science.

Biotechnology may be a very recent technological arena covering wide developmental studies in biological, agricultural, medical, and pharmacological sciences. Biomaterials are mostly soft and elastic, and rubber science is expected to play an important role in this arena, too. The more recent comer is environmental science. Environmental issues were as old as the history of mankind, but one of its beginnings as a branch of modern science is assumed to be the publication of "*Silent Spring*" in 1962, authored by Rachel Carson (1907–1964), who warned the wide environmental pollution by synthetic chemicals [28]. In Japan, the rapid industrial development in the 1950s and the 1960s often damaged the environments

Table 1.2 New technology and its object

Materials science	Material
Information science	Information
Biotechnology	Life-related technique
Environmental science	Environment

so much that in many industrial cities atmospheric air and environmental water (sea, lake, and river) were polluted to give damages to the healthy life of the inhabitants. Having recognized these negative factors of civilization, the movement involving inhabitants and scholars has pressured to create a new science, i.e., environmental science particularly in Japan [29]. Unhappy globalization of these pollution problems has accelerated the necessity of environmental science all over the world. It is still establishing itself, but from the SD viewpoint, its further progress might be uniquely crucial for the future of us.

Besides the ones listed in Table 1.2, there have appeared lots of new technologies, since the last quarter of the twentieth century. Many of them are more or less modeled after the engineering science, but have encountered a difficulty in systematizing and rationalizing themselves into the unified format from a theoretical viewpoint. Even the cybernetics, which is the oldest among them and has once enjoyed the strongest support as a novel promising scientific discipline, seems to have been absorbed into information science and biotechnology at present.

On the innovation of rubber science, too, there have been much efforts paid by many rubber technologists. Still, we have much to do for the sake of its establishment as a branch of not the traditional but of the most modern science. Rubber as a matter was turned out to be a material of much use by the invention of vulcanization in 1839, which was achieved by C. Goodyear. And the industrial manufacturing of bicycle and automobile rubber tires from NR began during the last decade of the nineteenth century, and it was established as a modern industry in the first quarter of the twentieth century, while the science of rubber was not established yet. The establishment of the macromolecular theory by H. Staudinger was in the 1930s, most probably around 1940, particularly due to the classical rubber elasticity theory (see Sects. 2.3.2 and 2.3.4). These historical facts seem to have impressed a seal of "traditional" on the science and technology of rubber as a branch of polymer science. We, rubber scientists or engineers, have to take lessons from the progress in chemistry: During the secession of several scientific disciplines such as geology, botany, geography, and anthropology into modern sciences from the traditional natural history, chemistry was establishing itself as a branch of modern science on the basis of the return of ancient atomic theory [30], with absorbing the alchemy of the Middle Age [31]. Of course, during the process of establishing chemistry as a modern science, extraordinary efforts by many scientists, such as R. Boyle (1627–91), J. Black (1728–99), J. Priestly (1733–1804), A. Lavoisier (1743–94), J. Dalton (1766–1844), and J. Berzelius (1779–1848), to name several, are surely to be noted [32–34].

The traditional "rubber science and technology" has to be innovated by adapting more modern techniques in chemistry and physics for its future progress while maintaining the essence of the traditional scientific aspects of rubber technology. The present text is intended to contribute to the progress toward the modernization of rubber science, which is to be discussed again later in Sect. 1.3.3 together with the discussions on what are technics and technology in general.

1.2 Natural Rubber: A Unique Biopolymer

1.2.1 Characteristics of Natural Rubber

Natural rubber (NR), the main chemical component of which is *cis*-1,4-polyisoprene, is a biomass; hence, it is renewable and carbon neutral, i.e., it is biosynthesized from carbon dioxide as a carbon source, and does not contribute to the increase of it in the atmosphere [21, 35–38]. Among lot of biomasses, NR is now enjoying one of the most wide industrial uses as a soft and tough material. Its unique qualities may be summarized [38] as follows.

Firstly, NR is biosynthesized in *Hevea* trees (*Hevea brasiliensis*), and the exudate, i.e., latex containing NR is collected after the incision of the trunk. These operations are called tapping, which is manually done using a special knife to give a cut on the tree trunk, a skillful operation needing a special proficiency [21, 37]. By the way, the *Hevea* tree was once named as the "Para rubber tree" which was originated from the Para port situated at the Amazon River just before flowing into the Atlantic Ocean. NR tapped at the Amazon Valley was exported via the port to the world. And the name is still widely used among botanists and agricultural scientists. Interestingly, the situation is very similar to that of "Mocha coffee": African coffee was exported through the Mocha port at the Arabian Peninsula to be known as Mocha coffee. Return to the first point, however, the fact that NRs being collected from a single species, *Hevea brasiliensis*, is recently considered to be a disadvantage from the viewpoint of biodiversity and biosecurity [21, 38–41]. We are going to face this problem of NR in a near future.

Secondly, from the nineteenth century on, the synthetic organic chemists have been successfully synthesized many natural products. However, the chemical (not biochemical) synthesis of NR has not done yet. One of the reasons is the highest stereo-regularity of *cis*-1,4-polyisoprene in NR: It contains only two or three *trans*-1,4-monomeric units at a terminal end of the linear polymeric chain. When we assume 3 *trans* units in a polymeric rubber chain, the degree of polymerization of which is one thousand (a relatively low value of NR), the ratio of units other than *cis*-1,4 is only 0.3%. Additionally, the three *trans* units are situated at one of the terminal ends of the chain. Such a stereo-regular *cis*-1,4-polyisoprene has not chemically synthesized yet [21, 39]. This high regularity gives rise to the specific feature of NR, i.e., unique strain-induced crystallization (SIC) behavior of NR (see Sect. 3.4), which is of utmost importance in evaluating most physical properties of NR. SIC is in fact a corollary of the high stereo-regularity of NR.

The tread rubber of tires for aircraft, which is of course an organic substance, is exposed to the harshest conditions when taking off and landing in terms of mechanical dynamic stress and friction (hence very high temperature) during the dynamic contact with the road surface of the air field. Yet, rubber tires, cap tread rubber of tire (see Chap. 5 for tread rubber) in particular, have managed to enable safe movements of aircraft on the surface of the earth, while arresting the destruction of them. SIC is regarded to be one of the responsible characters of NR for usages under so harsh a condition, together with wearing/abrasion properties of NR [38–41].

Thirdly, the elasticity of rubber is uniquely originated from entropy [21, 42]. The elasticity observed in usual matters is thermodynamically an energy originated. Other than rubber, inorganic glass is also amorphous (not crystalline, see Sect. 1.3), but its low elasticity is due to energy. In terms of elasticity, rubber has close relation with gas, the elasticity of which is entropic, too. The pneumatic rubber tires are a uniquely invented product by putting rubber and air together, to have enabled the automobiles to be driven in safety on the roads, not on the rails [21].

From these characteristics, rubber is considered to be a material of utmost importance for the sustainable development of the modern transportation society. It is fortunate for human beings to have found NR in the natural world, before the invention of synthetic rubbers. Also from the chemists' point of view, it is challenging that we have not yet succeeded in chemical synthesis of this natural product, NR.

1.2.2 Synthetic Natural Rubber?

NR is biosynthesized in a *Hevea* tree, and the tapping of the latex-bearing cortex of *Hevea* trees makes NR latex to exude through the incision. The latex contains the solid NR in the so-called rubber particle, and we get the solid-state NR by casting the latex onto a plate to obtain a NR thin film after evaporating the dispersant, or by coagulating the latex using aqueous acidic solution to obtain a NR bulk. Since the 2nd half of the nineteenth century, many organic chemists have challenged to the chemical synthesis of NR. However, the perfect chemical synthesis is still to be investigated in this century. Styrene-butadiene rubber (SBR) was developed in the 1930s and was industrially manufactured during the WW II. It is still mass-produced for tire rubbers of passenger cars and so on, and is one of the main rubbers for general purposes now. The chemical structure of it is shown in Fig. 1.2 with that of NR.

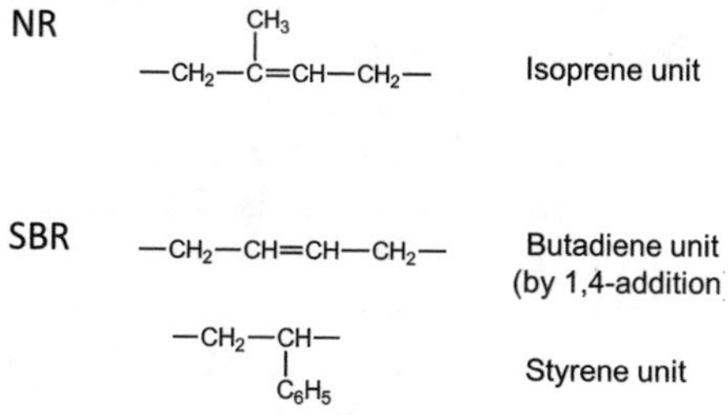

Fig. 1.2 Chemical structure of NR and SBR

NR $-CH_2-\underset{\underset{CH_3}{|}}{C}=CH-CH_2-$ Isoprene unit

SBR $-CH_2-CH=CH-CH_2-$ Butadiene unit (by 1,4-addition

$-CH_2-\underset{\underset{C_6H_5}{|}}{CH}-$ Styrene unit

In Fig. 1.2, the isoprene monomeric unit of NR is in *cis* configuration, and the chemical name of NR is *cis*-1,4-polyisoprene. SBR is a random copolymer of styrene and 1,3-butadiene (here only 1,4-addition unit is shown). Many grades of SBR are available in the market, and their properties are dependent on the content of styrene units, addition modes of 1,3-butadiene (1,2 or 1,4), *cis* or *trans* configuration of the C=C double bond, to result in covering a wide range of properties. The most popular one is manufactured by emulsion polymerization technique with styrene content of 23.5%, which is a general-purpose rubber as popular as NR (see the table in "Appendix").

Isoprene rubber (IR) was developed in the 1960s, which has almost the same chemical structure as that of NR shown in Fig. 1.2. It is uniquely called "synthetic natural rubber". Some of its properties are more or less similar to those of NR, but the stereo-regularity of IR is between 95 and 98% of *cis*-1,4, compared with almost 100% of NR. This apparently "small" difference in the stereo-regularity results in the difference between NR and IR: Several important physical properties of IR are not as good as NR, mainly due to less tendency for strain-induced crystallization (SIC, see Sect. 3.4.2). IR still remains to be a substitute of NR so far.

1.3 Rubber and Elastomer as Amorphous Polymers

1.3.1 Amorphous

Different from a crystalline state, the amorphous does not have structural regularity at the nanometer scale [43]. Gas and liquid, except liquid crystal, are amorphous. Exactly as Ziman claimed generally in his book [44] on polymeric materials, "few macromolecules form regular crystalline states, and they display complicated topologically randomized states to give rise to difficulty in analytically describing the state of them." As a matter of facts, it may not be easy to differentiate the assembly of fine crystallites and/or the para-crystals (crystals with lots of defects) from the amorphous state of polymers without using X-ray diffraction analysis. Specifically, the wide-angle X-ray diffraction (WAXD) pattern displays isotropic clear ring, and the orientation of the crystallites is random, which is the case for spherulitic crystals (see Sect. 3.4.3). When the pattern shows isotropic ring-shaped halo, the amorphous state is proved (see Sect. 3.4.1). On an amorphous phase, the random network model by Zachariasen [43–46] has been well known, and the most suitable example of it may be the rubber. For example, SBR is a random copolymer of styrene and 1,3-butadiene, and is not amenable to crystallize due to the structural randomness at nanometer scale. Additionally, even stereo-regular NR is also amorphous due to the active micro-Brownian movement of the polymeric chains. In the case of NR, however, it may crystallize (low-temperature crystallization) if kept standing at lower temperature for a long time. Also, the cross-linked NR crystallizes upon stretching, which is called strain-induced crystallization (SIC), and SIC is the origin of unique high performance of NR products. Note that SIC is reversible on stretching and contracting, and they are amorphous unless under a high strain (see Sect. 3.4 for the details of NR crystallization).

From a structural standpoint, amorphous states do not have a solid–solid interface due to the absence of a crystalline region, and hence, they apparently afford the homogeneous matrix at least macroscopically. This is a base of the statement, "Rubber is a polymeric solvent, though its viscosity is high." This liquid-like amorphous property of rubber affords much accommodating ability to rubbery matrix, enabling mixture with various powdery solids and liquids including various fillers and many reagents, ultimately soft composites of superior properties.

1.3.2   Glass Transition Temperature

Solid-to-liquid transition occurs when the crystalline materials are heated to their melting temperature (T_m). Amorphous means the absence of melting, but the rubber is soft and more or less liquid-like at room temperature. It is apparently a solid, but such appearance is due to its very high viscosity, and years of standing may display its liquid-like nature (cold flow). Combined this liquid-like nature of rubber with the amorphous nature of rubber, the rubbery matrix is known as a polymeric solvent, which can accommodate various additives, for example, carbon black as a reinforcing filler. When brought to a lower temperature than its glass transition temperature (T_g), it is exactly an amorphous solid. Heating of it to above T_g turns it to a soft polymeric solvent again (see later Sect. 2.4.1 for more detailed discussion on T_g).

Any polymeric materials whose T_g is lower than room temperature are a candidate for rubber. Therefore, T_g of rubber is somewhat equivalent to T_m of crystalline materials. Commercially available rubbers are listed in the table in "Appendix" with their specific features. Their lowest usable temperatures are closely related to their T_g's, all of which are below room temperature. Thus, T_g is an industrially very important index of rubber for practical usages. For example, so-called snow tires are made from the rubber the T_g of which is well lower than -30 °C to make them workable on snow or ice.

1.3.3   The Age of Soft Materials and Soft Technology

The twenty-first century, when the Iron Age seems to be just passing its zenith, may be the century of seeing the start of Age of Soft Materials. Rubbers are surely one of the earliest items to be recognized its utility among soft materials, which suggests us a reason to study rubber science in this century. Rubbers, polymer gels, colloids, and liquid crystals are expanding their utilization in accordance with the modern technological trend of devices that are soft, flexible, lightweight, and still being of tenacity. For example, the robots for industrial uses and in particular those for uses under dangerous conditions in emergency are already recognized their usefulness. The further development direction of the robotics may recently be suggested, the robots of much more humane characters. Next-generation robots are the ones that people want to hold in his or her hands or to be embraced by them. To realize such a humane character, it is mandatory to develop the devices with soft touch of the surface, highly deformable muscles, and flexible joints and so on.

As shown in Fig. 1.1, when we use them not as a material but as a matter, the Age of Soft Matters is the name to use from a basic scientific viewpoint. Also, this century may be that of entropy, while the twentieth century was that of energy. Entropy is to determine not only the elasticity of rubber and gas but also the quality of energy. The second law of thermodynamics is a principle of increasing entropy, and in an isolated system, the state is simply tending to be randomized toward chaos, which is measured by the increase of entropy [47, 48]. By our consumption of energy, the quality of energy is decreasing on the earth, i.e., the entropy of the system is constantly increasing. We are now not in energy crisis but in entropy crisis! The significance of studying rubber science includes possibly our understanding entropy more. From the viewpoint of materials science, the present state of our science on rubber and elastomer is not advanced enough to surpass the limit of a trial-and-error methodology. By publishing this textbook, we have tried to summarize the technical results on rubber which have been accumulated in the last century, and to advance one step toward the establishment of modern rubber science, based on the methodology in sciences of soft matters and of entropy.

Having this long picture in mind, let us learn a little more on technology. Actually, the term "technology" has now dual meanings; it means either technic itself or science on technics. In this chapter and later in Chap. 6, however, technology is employed for science on technics: Originally, "-logy" means learning or theory and technology is conceptually a branch of science. Therefore, often used phrase "science and technology" is problematic, and "science and technics" is recommendable from the present viewpoint. On this standpoint, what is the technic is considered first. Its origin is a few tools that our ancestors used for working on nature. That is, the mankind began to separate from the other species by using some tools, not directly using his or her hands on the natural world in order to work for survival [20]. Their use of tools had changed the nature in a long time to form the environment around the place where the human beings were living. In other words, environment is a nature more or less changed by the human activities. Utilization of tools had promoted the evolution of the human beings, and the progress of simple tools into various machines ultimately enabled mankind to industrialize our society. The result of these considerations on technics is schematically shown in Fig. 1.3.

The humankind had possibly used a chipped stone tool when working on nature for their survival, and such activities of them later resulted in invention of the polished stone tools, etc, were followed by the gradual conversion of nature to environment. In a sense, the environments are the artificial nature. During the Stone Age, human being learned how to use fires, sophisticated stone tools, and began even metals by chance for their survival and to progress civilization. This is the

Fig. 1.3 Technics are intermediating between human beings and environment, and all are within nature in its broad sense

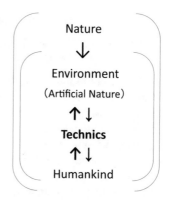

most important difference of humans from all the other living things. Actually, the use of a tool, an origin of technical instruments and later evolved into machines, was the accelerating factor for the ape-men to evolve into human beings, and the usage of technics by the humankind in working on nature had given rise to civilization. L.H. Morgan (1818–1881), an early anthropologist in America, who had conducted extensive sociological field research on native American Indians, wrote in the first chapter of his book "Ancient Society" [49] as follows:

> The latest investigations respecting the early condition of the human race, are tending to the conclusion that mankind commenced their career at the bottom of the scale and worked their way up from savagery to civilization through the slow accumulations of experimental knowledge.

Here, "experimental knowledge" is reasonably interpreted as the primitive technics on manufacturing and effective utilization of various tools (initially stones and woods) accumulated through their activities such as hunting, gathering, and culturing. That is, the technics, however primitive they might be, is the driving force for civilization [50].

Nowadays, we are encircled not by the nature but by the environment. The progress of mankind has been accelerating and so is the global widening of environment. Through the Industrial Revolution [4, 20, 51], in particular, the development and worldwide spread of modern technics were much accelerated, and the trend has not diminished until now [40, 52–54]. Recent international race to the Red Planet, Mars, is even leading to be the human's next big mission to colonize Mars [55]. In other words, our environment is even expanding to the outsides of the Earth! B. Russell (1872–1970), a famous English mathematician and philosopher, wrote in 1955 as shown below which appeared at the top of his chapter entitled "Science and Human Life" of the book [56]:

> Science and the techniques to which it has given rise have changed human life during the last hundred and fifty years more than it had been changed since men took to agriculture, and the changes that are being wrought by science continue at an increasing speed.

And Russell closed his chapter by the following sentence:

No previous age has been fraught with problems so momentous; and it is to science that we
must look for a happy issue.

At present in the twenty-first century, however, the progress of technics is to be destined to follow the sustainable development (for more details of SD, see Chap. 6), which is now the most urgent issue for our future. The role of modern technology on state-of-the-art technics, a branch of science, is more and more indispensable for the survival of human beings on the earth. Naturally, we who are studying and working on rubber are responsible for the progress of rubber science, in order to contribute to solving the problems we are facing. For these assignments on human beings, understanding the counter-reactions from nature and environment shown by the downward arrows in Fig. 1.3 is becoming more and more important in order to make technics and hence technology more flexible and much more friendly toward humanity.

Remark 1 The Properties of Commercialized Rubbers
Nowadays, many kinds of rubber are commercially available. When starting a research on rubber, choosing one of them is an idea. Also, it is beneficial or even mandatory to have some knowledge on them for understanding rubber science in general. Since this text has failed to describe all of them, a brief survey of their properties is given here. At first, look at the table in "Appendix". The table lists various properties showing their characteristics together with their chemical structures. Thermoplastic elastomers (TPE, see Sects. 3.5.3 and 3.5.4 for TPE) are not included in the table. For more information on each rubber, request qualification sheets or catalogs from the providers.

Natural rubber (NR) and isoprene rubber (IR, also called synthetic natural rubber) are among the general purpose rubbers used in many rubber products including tires and rubber belts. Styrene-butadiene rubber (SBR) is a random copolymer of styrene and 1,3-butadiene by emulsion polymerization, and is a typical synthetic general purpose rubber. Its tendency to crystallize is the lowest, to be a representative amorphous material. Since the latter half of the 1980s, SBR by solution polymerization has become popular together with butadiene rubber (BR) for high-performance tires. Nitrile rubber (NBR) is a copolymer of acrylonitrile and 1,3-butadiene, and is of high oil resistance. (HNBR, a hydrogenated one is more heat resistant than NBR.) It is a typical example of specific purpose rubbers. Chloroprene rubber (CR) is the first commercialized synthetic rubber, and it seems to be an all-rounder of value due to its excellent mechanical properties. Butyl rubber (IIR) is a copolymer of isobutylene and 1,3-isoprene, and is of use for inner tube of pneumatic tires because of its low gas permeability. (BIIR, brominated IIR, is also popular for the tube.) All that mentioned in this paragraph form a group, diene rubbers, and their acronyms end by letter R.

Ethylene–propylene rubbers (EPM, EPDM) are a copolymer of ethylene and propylene and a terpolymer of the two monomers plus a diene compound (for enabling sulfur cross-linking), respectively. They are of value due to their high heat resistance and good weatherability, and are all-rounders second to CR. Ethylene–vinyl acetate copolymer (EAM), chlorinated polyethylene (CM), chlorosulfonated polyethylene (CSM), acrylic rubber (ACM, copolymers of ethyl- or butylacrylate) are all special purpose rubbers for their specific properties, and their acronyms end by M.

Additionally, hydrin rubbers (CO, ECO), urethane rubber (U), polysulfide rubber (T) have to be mentioned as the rubbers for specific functional purposes. Silicone rubber (Q) and fluoro rubber (FKM) are most unique rubbers, and an account of them is given in **Remark 3**. The following table may explain the last letter in the acronym of rubber which is used in the ASTM naming of rubbers.

Table	Generic nomenclature of synthetic rubbers
M	Saturated poly- or oligo-methylene chain
N	Nitrogen on the polymer chain
O	Carbon and oxygen in the polymer chain
R	Unsaturated carbon chain
Q	Silicon and oxygen in the polymer chain
T	Carbon, oxygen, and sulfur in the polymer chain
U	Carbon, nitrogen, and oxygen in the polymer

References

1. S. Kohjiya, in *Shizen o Kangaeru (Think of Nature)*, ed. by K. Izumi, A. Sasabe, (Horitsu Bunkasha, Kyoto, 1987), pp. 135–155 [In Japanese]
2. T. Forester (ed.), *The Materials Revolution: Superconductors, New Materials, and the Japanese Challenge* (MIT Press, Cambridge, MA, 1988)
3. K. Geiser, *Materials Matter: Toward a Sustainable Materials Policy* (MIT Press, Cambridge, MA, 2001)
4. J.D. Bernal, *The Social Function of Science* (Routledge & Kegan Paul, London, 1939)
5. N. Campbell, *What is Science?* (Dover, New York, 1953)
6. P. Frank, *Philosophy of Science* (Prentice-Hall, Englewood Cliffs, 1957)
7. V.I. Lenin, in *V. I. Lenin Collected Works*, vol. 14 (Foreign Language Publishing House, Moscow, 1962) [The original was published in 1909]
8. J. Losee, *A Historical Introduction to the Philosophy of Science* (Oxford University Press, Oxford, 1972)
9. K.R. Popper, *The Logic of Scientific Discovery* (Routledge, London, 1992) [The original German edition was published in Vienna 1934. First English edition was published 1959 by Hutchinson Education]
10. J. Henry, *The Scientific Revolution and the Origin of Modern Science*, 3rd edn. (Palgrave Macmillan, Basingstoke, 2008)
11. C.H. Langmuir, W. Broecker, *How to Build a Habitable Planet: the Story of Earth from the Big Bang to Humankind* (Princeton University Press, Princeton, NJ, 2012)
12. J. Bronowski, *The Ascent of Man* (Little, Brown & Co., Boston, 1973)
13. A. Robert, *The Incredible Human Journey* (Bloomsbury Publishing, London, 2009)

14. P. Bellwood, *First Farmers: The Origin of Agricultural Societies* (Blackwell Publishing, Oxford, 2005)
15. L. Beck, *Die Geschichite des Eisens in Technischer und Kulturgeschichtlicher Beziehung, I-V* (Friedrich Vieweg und Sohn, Braunschweig, 1884–1903)
16. F. Dannemann, *Die Naturwissenschaften in ihrer Entwicklung und Zusammenhange* (Die Zweite Auflage in Vier Bänden, Leipzig, 1920–1923)
17. S. Mason, *A History of the Sciences: Main Currents of Scientific Thought* (Abelard-Schuman, London, 1953)
18. J.D. Bernal, *Science in History* (C. A. Waltts & Co., London, 1965)
19. H. Hodges, *Technology in the Ancient World* (Penguin Press, London, 1970)
20. T.K. Derry, T.I. Williams, *A Short History of Technology: From the Earliest Times to A. D. 1900* (Dover, New York, 1993) [The original edition was published in 1960 by Oxford University Press]
21. S. Kohjiya, *Natural Rubber: From the Odyssey of the Hevea Tree to the Transportation Age* (Smithers Rapra, Shrewsbury, 2015)
22. E.D. Larson, M.H. Ross, R.H. Williams, in *Materials Revolution*, ed. by T. Forester (MIT Press, Cambridge, MA, 1988), pp. 141
23. G.S. Whitby, C.C. Davis, R.F. Dunbrook (eds.), *Synthetic Rubber* (Wiley, New York, 1954)
24. J.P. Kennedy, E.G.M. Törnqvist (eds.), *Polymer Chemistry of Synthetic Elastomers* (Interscience Publishers, New York, 1968)
25. S. Kohjiya, Gomu no Rekishi, in *Gomu no Jiten* (*Encyclopedia of Rubber*), ed. by M. Okuyama, S. Kohjiya, T. Nishi, K. Yamaguchi (Asakura Shoten, Tokyo, 2000), Chap. 1 [In Japanese]
26. E. Zilsel, Am. J. Sociol. **47**, 544 (1942)
27. P.-M. Schuhl, *Machinisme et Philosophie, la troisième édition* (Presses Universitaires de France, Paris, 1969)
28. R. Carson, *Silent Spring* (Houghton Mifflin, New York, 1962) [Reprinted in Penguin Classics in 2000, in which the fascinating and appealing pictures in the original are regrettably excluded]
29. H. Shoji, K. Miyamoto, *Osorubeki Kogai* (*Terrible Pollution Problems in Japan*) (Iwanami Shoten, Tokyo, 1964) [In Japanese]
30. S. Greenblatt, *The SWERVE: How the World Became Modern* (W. W. Norton, New York, 2011)
31. H.M. Leicester, *The Historical Background of Chemistry* (Wiley, New York, 1956) [Reprinted by Dover in 1971]
32. I. Freund, *The Study of Chemical Composition: An Account of its Method and Historical Development* (Cambridge University Press, Cambridge, 1904) [Republished from Dover in 1968]
33. J.R. Partington, *A Short History of Chemistry* (Dover, New York, 1989) [This Dover edition was the revised 3rd one. Original one was published in 1937 by Macmillan & Co]
34. J.-P. Poirier, *Lavoisier: Chemist, Biologist, Economist* (University of Pennsylvania Press, Philadelphia, 1996)
35. L. Bateman (ed.), *The Chemistry and Physics of Rubber-Like Substances* (MacLaren & Sons, London, 1963)
36. A.D. Roberts (ed.), *Natural Rubber Science and Technology* (Oxford University Press, Oxford, 1988)
37. C.C. Webster, W.J. Baulkwill (eds.), *Rubber* (Longman, Harlow, 1989)
38. S. Kohjiya, Y. Ikeda (eds.), *Chemistry, Manufacture and Applications of Natural Rubber* (Woodhead/Elsevier, Cambridge, 2014)
39. Y. Ikeda, A. Tohsan, S. Kohjiya, in *Sustainable Development: Processes, Challenges and Prospects*, ed. by D. Reyes (Nova Science Publishers, New York, 2015), Chap. 3
40. A.L. Tullo, Chem. Eng. News, April 20, 18 (2015)
41. Y. Ikeda, P. Junkong, T. Ohashi, T. Phakkeeree, Y. Sakaki, A. Tohsan, S. Kohjiya, K. Cornish, RSC Advances **6**, 95601 (2016)

42. L.R.G. Treloar, *The Physics of Rubber Elasticity*, 3rd edn. (Clarendon, Oxford, 1975)
43. S.R. Elliott, *Physics of Amorphous Materials*, 2nd edn. (Longman, Harlow, 1990)
44. J.M. Ziman, *Models of Disorder* (Cambridge University Press, Cambridge, 1979)
45. W.H. Zachariasen, J. Am. Chem. Soc. **54**, 3841 (1932)
46. A.R. Cooper, J. Non-Cryst. Solids **49**, 1 (1982)
47. J.D. Fast, *Entropy: The Significance of the Concept of Entropy and its Applications in Science and Technology*, 2nd edn. (Macmillan, London, 1968)
48. P.W. Atkins, *The Second Law* (Scientific American Books, New York, 1984)
49. L.H. Morgan, *Ancient Society* (Henry Holt Co., New York, 1877) [Based on the results of Morgan's field work described in this book, F. Engels authored his book 50, Ch. 1 of which is a summing up of the Morgan's field works]
50. F. Engels, *The Origin of the Family, Private Property and the State* (Penguin Classics, London, 2010) [The original was published in 1884]
51. T.S. Ashton, *The Industrial Revolution, 1760–1830* (Greenwood Press, Westport, 1986)
52. W.H.G. Armytage, *A Social History of Engineering* (Faber and Faber, London, 1961)
53. L. Turner, W.C. Clark, R.W. Kates (eds.), *The Earth as Transformed by Human Action: Global and Regional Changes in the Biosphere over the Past 300 years* (Cambridge University Press, Cambridge, 1990)
54. D. Arnold, *The Problem of Nature: Environment, Culture and European Expansion* (Blackwell Publishers, Oxford, 1996)
55. J. Achenbach, *National Geographic* (2016) [November, 30, with the special poster Colonizing Mars]
56. J.R. Newman (ed.), *What is Science?* (Simon and Schuster, New York, 1955) [Republished in 1961 by Washington Square Press, New York]

Chapter 2
Basic Science of Rubber

2.1 Chemistry I: Polymerization, Polymer Reaction, and In Situ Chemical Reaction

2.1.1 Polymerization: Synthetic Rubbers

(1) Polymerization Reactions

Polymerization is a chemical reaction to synthesize polymers (high molar mass substances) from monomers (low molar mass ones). Originally, polymerization was for addition type chain reactions [1], the reactive intermediate of which is free radical or ionic intermediates (cation and anion). Nowadays, however, condensation-type reactions toward polymers using, e.g., di-alcohols, di-carboxylic acids, and di-amines are also classified into polymerization. Another type reaction, polyaddition is for urethane elastomers. Polyether and polyester prepolymers for polyurethanes are synthesized by ring-opening polymerization.

Biosynthesis of natural rubber may also be a kind of polymerization, but biosynthesis is classified independently due to much difference from polymerization. Methyl rubber, which was the first synthetic rubber industrially manufactured before and during the World War I, was prepared by heat polymerization [2, 3]. It might be equivalent to free radical polymerization in the present scientific terminology.

In terms of kinetics, polymerization is a typical complex reaction, and its elementary reactions include initiation, propagation, transfer, and termination reactions [4–6]. In order to elucidate the polymerization mechanisms, structural determinations (both chemical and the higher order structures) of the reaction products have been a powerful tool as well as the kinetic study of the reactions [7, 8].

(2) Radical Polymerization

Carbon-free radical (–C·), which is electrically neutral, is the reactive intermediate in the radical polymerization. Several synthetic rubbers (SBR, BR, NBR, etc.) are

Y. Ikeda et al., *Rubber Science*, https://doi.org/10.1007/978-981-10-2938-7_2

manufactured by emulsion or solution polymerization technique using a free radical initiator [1, 7, 8]. In addition, the cross-linking reaction of rubber by peroxide, auto-oxidation of rubber by atmospheric oxygen and ozonolysis of rubber (decomposition of carbon-carbon double bonds in the rubber molecule by atmospheric ozone) are by the free radical reaction mechanism in the rubber arena. Therefore, to understand radical reactions is of importance in studying rubber.

Industrial preparation of synthetic rubbers was first started by radical polymerizations as mentioned above. In 1930s, radical copolymerization of styrene and 1,3-butadiene for SBR and radical polymerization of chloroprene for CR was developed in Germany and in the USA, respectively [7, 8]. Radical polymerization is highly exothermic (accompanied by the heat generation), and emulsion polymerization (using an aqueous medium) was chosen for the easier temperature control. In the aqueous medium, hydrophobic monomers (styrene and 1,3-butadiene) are dissolved into micelle, i.e., emulsification of monomers being precondition, and a free radical segment produced by the unimolecular decomposition of initiator comes into the micelle to initiate the polymerization of the two monomers, i.e., copolymerization takes place inside the micelle. The product is an emulsion containing SBR in the micelles. SBR still remains to be a major mass-produced general-purpose rubber in this century. A few grades of BR and NBR are manufactured by emulsion polymerization, and EAM and ACM are manufactured by radical polymerization, too. On these rubbers cited by a shortened symbol, refer to Remark 1 and "Appendix".

(3) **Cationic, Anionic, and Coordinated Anionic Polymerizations**

In cationic and anionic polymerizations, the reactive intermediates are carbocation ($-C^+$) and carbanion ($-C^-$), respectively. In these ionic polymerizations, the reactive intermediates are ordinarily present in a form of ion pair in accordance with the electroneutrality principle, that is, $-C^+ A^-$ in cationic polymerization and $-C^- B^+$ in anionic polymerization, where A^- is a counter anion and B^+ is a counter cation, respectively. Under a certain condition especially in a polar solvent, the ion pair may be dissociated into free ions. Both polymeric free carbocation and free carbanion are much more reactive than their paired forms. Hence, the choice of solvent is an important factor for ionic polymerization. In the selection, solubility of the produced polymer has to be taken into account, too.

Butyl rubber (IIR, isobutylene-isoprene copolymer) is manufactured by low-temperature cationic polymerization [8]. In general, the carbocation is less stable (hence, more reactive) than the carbanion. Thus, the control of the side-reactions such as chain transfer is practiced by conducting the reaction at a very low temperature, e.g., at -70 °C or at a lower temperature, in order to obtain the linear IIR of as high a molar mass as possible.

At an early stage of synthetic rubber development in Germany and in Russia, it was found that a metallic sodium (Natrium in German), which was used for the dehydration of monomers and non-polar organic solvents, was found to initiate the polymerization of 1,3-butadiene, which is an early example of anionic polymerization. From these historical facts, Buna (<u>Buta</u>dien-<u>Na</u>trium) had been a word used for synthetic rubber in Germany. Later, SBR by emulsion polymerization was named Buna-S in Germany. In 1950s, living anionic polymerization was established by using metallic sodium or butyl lithium (BuLi) as an initiator, and living polymerizations (without termination) of styrene, 1,3-butadiene, and isoprene were carried out [8, 9]. As already mentioned, carbanion is relatively stable compared with carbocation and by suitable purifications of all the reagents the termination reaction of the carbanion was eliminated, to result in the living polymerization of these monomers. Living polymerization of isoprene by BuLi afforded *cis*-1,4-polyisoprene (IR, isoprene rubber), the *cis*-1,4 content of which was over 90%. IR had an enigmatic name "synthetic natural rubber", since its stereo-regularity was relatively high. However, as aforementioned, its properties are inferior to NR due to less stereo-regularity than NR whose regularity is almost perfect.

In the 1980s on, many grades of livingly polymerized (not by an emulsion but by a solution technique of anionic mechanism) BR and SBR have been developed, and applied to lots of rubber goods including passenger car tires. The two are designated as s-BR and s-SBR, and the emulsion polymerized ones are now designated as e-BR and e-SBR, respectively. Solution polymerization has enabled various microstructures and molar mass to cover wider ranges of physical properties. Also, taking advantages of living polymerization, introduction of a chemically functional group at the rubber chain ends has been conducted successfully (see Sects. 2.2.2 and 5.4.2). The functional group at the chain ends is useful in controlling interaction with the reinforcing fillers to result in higher performance rubber. Thus, the demands for s-BR and s-SBR are increasing still now.

The most impressive application of living anionic polymerization is the development of new thermoplastic elastomers, SBS and SIS: Both are ABA type triblock copolymers consisting of three blocks:

Polystyrene-Polybutadiene (or Polyisoprene)-Polystyrene

where polystyrene is A component, and polybutadiene in SBS and polyisoprene in SIS are B component. The molar mass of each block is easily controlled since the living polymerization technique is used. The marketing of SBS and SIS was a chiming of the bell to announce a coming of new rubber, thermoplastic elastomer (TPE, see Sect. 3.5.3), which can be usable without vulcanization [10].

K. Ziegler (1898–1973), a German chemist working on organometallic compounds, had blazed a trail for a series of new organometallic compounds, by which he found ethylene was polymerized smoothly to polyethylene of very high molar mass. Until then, ethylene had been known difficult to be polymerized, and the polyethylene was obtained only under the high pressure. His catalysts, later named Ziegler catalyst, pioneered a new stage of polymer chemistry by establishing the polymerization of ethylene to PE, propylene to PP, and several other olefins. He received the Nobel Prize in Chemistry in 1963. G. Natta (1903–1979) conducted structural analysis of PP, which was found crystalline. He elucidated the optical isomerism of PP due to its stereo-regularity, and proposed two types of regularity, i.e., isotactic and syndiotactic. He was the cowinner of the Nobel Prize with Ziegler. The Ziegler catalyst is heterogeneous, that is, not soluble in organic solvents. Hence, the coordination of propylene monomer to the active anionic site was reasonably assumed to be important for the isotactic or syndiotactic polymer synthesis. The polymerization by Ziegler and the analogous organometallic catalysts is named coordinated anionic polymerization, and the catalysts have been developed extensively for stereo-regular polymerization of various monomers [9, 10]. One example is IR of 96–98% cis-1,4 content. In terms of regularity, it is higher than IR synthesized by BuLi, though still lower than NR.

Ziegler catalyst for PP is active for copolymerization of ethylene and propylene to produce ethylene–propylene rubber (EPM), and for terpolymerization of the two monomers and a di-olefin to result in ethylene-propylene-diene terpolymer (EPDM), which is cross-linkable either by peroxide or by sulfur/accelerator system [8]. Together with the several other polyolefin rubbers on the market, they are now in wide use for their superior antiozone and antioxidation properties. See the table in "Appendix" for EPDM.

(4) New Polymerization Reactions

After the development of Ziegler catalysts, new synthetic rubbers (often called "elastomers") have been marketed to provide rubber technologists with various types of rubbers other than diene rubbers. Diene rubbers are the rubbers from diene monomers (1,3-butadiene or isoprene), including copolymers such as SBR and NBR. Additionally, developmental research on new organometallic compounds for polymerization reactions in general was focused among synthetic polymer chemists. Among them, metallocene-type catalysts of zircon (or titan) in combination with methyl aluminoxane, which were reported by Kaminsky [11], have much developed for the polymerization of olefins to give various elastomeric polyolefins [12]. It is reported that a catalyst system from the gadolinium metallocene has afforded highly stereo-regular cis-1,4-polybutadiene [13], which may possibly be of use for the tread rubber of tires.

The rapid progress of living radical polymerizations [14–17] is promising to provide us with new higher performance polymers for the rubber industry in a near future. Also, metathesis polymerization is potentially to afford olefinic elastomers even from noncyclic diene compounds [18].

Still to be solved is the total chemical synthesis of NR, which was already mentioned in Sect. 1.2.2. The almost perfect regularity has been via biochemical reaction by the enzyme. Thus, biosynthesis of NR in vitro might be the other possible target in a near future.

2.1.2 Polymer Reaction: Chemical Modification of Rubber and Elastomer

The polymer reaction is defined as any reactions on a polymer substrate, which surely includes vulcanization or any other cross-linking reactions of rubber. However, the cross-linking reactions are to be described in Sect. 2.3, considering their prime importance for rubber. In this limited meaning of polymer reactions of NR and elastomers, a few reviews have been published already [19–22]. In this section, therefore, the objective of polymer reaction is limited to synthesize a new rubber without using polymerization reactions. For example, hydrogenation of NBR (copolymer of butadiene and acrylonitrile) gives HNBR (hydrogenated NBR) which is actually a kind of terpolymer of ethylene, butadiene, and acrylonitrile since the perfect hydrogenation has not been done. HNBR is durable like EPDM as well as oil-resistant like NBR.

Polyethylene (PE) is simple in its chemical structure, and its low glass transition temperature (T_g) results in very flexible chains. This feature has suggested its potential utility as an elastomer, but it also tends to crystallize due to the simple chemical structure (see Sect. 3.4 for crystallization). Copolymerization has afforded an inhibitive action on the crystallization, to result in the marketing of EPDM. Polymer reactions, namely chlorination and chlorosulfonation, of PE gave CM and CSM, respectively, both of which displayed much better weatherability than the diene rubbers and have been on the market for long. (See the table in "Appendix" for the synthetic rubbers shown by the abbreviated symbols.) As mentioned in Sect. 2.1.1 (3), functional group introduction to the living anionic polymer chain end of s-BR and s-SBR is a polymer reaction. If the introduced functional group is selected so as to produce chemical interaction with reinforcing fillers like carbon black or silica, the higher performance soft composites are expected to be obtained. This type of functionalization has been developed in rubber industries to afford higher performance tires (see Chap. 5).

Chemical modifications of NR have a long history, and a lot of relevant papers have been published as cited in Refs. [19–22]. Yet, only a few examples have been industrially mass-produced so far. Hydrogenation of NR is much expected to improve its weatherability to produce an alternating copolymer of ethylene and propylene if the hydrogenation is quantitative. Also, many scientific papers on the grafting of methyl methacrylate (MMA) onto NR have been published. But, in neither case, a success in the rubber market has been limited yet: A few grades of PMMA-graft-NR did appear in the market, but were commercially not much successful. Another example is epoxidation of carbon-carbon double bonds in NR to give epoxidized NR (ENR), on which two short reviews [23, 24] are cited here. Two grades, ENR-25 and ENR-50 (the epoxidation percentages, 25 and 50%, respectively), are on the market. However, recently only ENR-50 is reported to be available per the request from the customers.

The technique, in which one or two of α-olefins other than ethylene are fed to the polymerization reactor of ethylene (or copolymerization of ethylene and propylene) using selected Ziegler or metallocene catalyst, has been industrialized to obtain new olefinic soft polymer blends: The products are often named "reactor polymer alloy" or "nascent alloy." If the reaction of polymerizing ethylene with the α-olefin has taken place in the reactor, they are a kind of polymer reaction. Related topics are reviewed in an article entitled "post polymerization modification" [25].

2.1.3 In Situ Chemical Reaction of Rubber

(1) In Situ Chemical Reaction

The Latin words "in situ" means "at place" or "on the spot". In situ chemical reaction is relatively a new idea in polymer chemistry, but it is a somewhat conventional one in rubber industry, although the detailed technical information has not much reported. As will be mentioned in Sect. 2.2, Goodyear, who invented vulcanization in 1839, did not recognize it as a chemical reaction. However, during the processing of rubber some chemical reactions may potentially take place at place, or they may inevitably occur at a certain step of the processing, since several reactive chemicals have been compounded for the vulcanization reactions. How to control or inhibit the possible chemical reactions at place before the vulcanization step, which are in situ chemical reactions of the negative effect, has been the issue of great concerns for rubber chemists and engineers. Namely, precuring reactions while processing the rubber before the final vulcanization step (called "scorch" in rubber industry) should be inhibited, even by using appropriate chemicals called scorch inhibitor or prevulcanization inhibitor (PVI). For this problem, also refer to Sects. 2.2 and 3.5.2. Here, a few recent examples of positive utilization of in situ chemical reaction are to be explained.

(2) Particulate Silica Generated In Situ in Rubbery Matrix

An in situ sol-gel reaction of tetraethoxysilane (TEOS) to generate silica in the cross-linked rubber matrix has been applied to the reinforcement of rubber. This process involves the swelling of rubber network using a silica precursor, e.g., TEOS, and subjecting its TEOS-swollen network to hydrolysis and condensation to generate in situ silica in the cross-linked rubbers. After the pioneering work on silicone elastomer reported by Mark et al. [26], Kohjiya, Ikeda, and their coworkers extensively studied the reinforcement of general-purpose diene rubbers by silica generated in situ, for example, styrene-butadiene rubber (SBR), acrylonitrile butadiene rubber (NBR), butadiene rubber (BR), epoxidized NR (ENR), isoprene rubber (IR), and so on. In 2000, these topics were summarized in a review paper [27], followed by a more recent one [28] and two chapters [29, 30].

In these cases, the rubber reinforcement effect of in situ silica was considerable, but the technique may be much limited: The introduction of silica has been conducted into rubber vulcanizates, i.e., on rubber products. For the production of

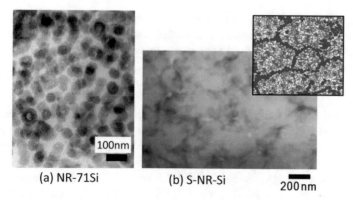

(a) NR-71Si   (b) S-NR-Si   200 nm

Fig. 2.1 Morphology of specifically controlled in situ silica-filled NR nanocomposites. **a** Homogeneously dispersed in situ silica in NR matrix (cited from Fig. 2 in Ref. [32]). **b** Filler network of in situ silica in NR matrix (cited from Fig. 1 in Ref. [36])

rubber goods of various shapes, this technique was introduced to uncross-linked rubber, on which the conventional processes using a milling and a pressing are to be utilized for preparing various shapes of in situ silica-filled cross-linked nanocomposites [28, 30–33]. Figure 2.1a shows the TEM photograph of the nanocomposite, where 71 phr of in situ silica in which diameter was 46 nm was filled [32]. The polarity and solubility of primary alkyl amine in water were the most influential factors for controlling the in situ silica content in the NR matrix. The speculated mechanism of the sol-gel reaction of TEOS in the swollen NR matrix was proposed using the generation of the reversed micelles, which was supported using high-resolution magic angle spinning NMR and hetero-nuclear correlation NMR measurements [34].

In addition, the in situ silica filling into rubbers was also conducted in rubber latexes [29, 30, 35–38]. The casting of the in situ silica-filled NR latex followed by latex curing was found to give a useful model of the network-like filler structure of silica generated in situ, where the rubber particle in NR latex functioned as a template to the filler network formation as shown in Fig. 2.1b [36, 37]. This template mechanism in the sol-gel reaction of TEOS is explained as follows: Rubber particles in the latex come closer during drying the latex (evaporation of water), and consequently the densely packed in situ silica is forced to disperse around the rubber particles, i.e., at the boundary layer between the surface of rubber particles and the surrounding aqueous phase.

A simultaneous synchrotron wide-angle X-ray diffraction (WAXD) and tensile measurements at a beamline in the SPring-8 elucidated a stepwise strain-induced crystallization (SIC) behaviors of the 10 phr in situ silica-filled NR nanocomposite [38], which was a different behavior from those of silica loaded one, onto which commercial silica was mechanically mixed. Peroxide cross-linked NR nanocomposites were similarly prepared by the soft processing method [30], where the in situ silica contents were 10 and 17 phr. The storage modulus (E') and loss modulus (E'') in the plateau regions of the samples were much higher than those of conventional-milled commercial silica-filled samples [37]. The results strongly suggested the important role of filler networks for reinforcement of rubber, which was achieved by using the in situ silica filling in the NR latex.

(3) Use of Coupling Agents

In rubber industry, it is not so common that coupling agents are mixed with rubber in advance, but the coupling agents are added at an adequate processing step and are reacted with rubber in situ at the final step of heat-pressing. In the system using bis [3-(triethoxysilyl) propyl] tetrasulfide (TESPT) [39] as the silane coupling agent, disulfidic linkage (–S–S–) in TESPT is reacted with rubber chains as well as sulfur (S_8), and the alkoxysilyl groups in TESPT are hydrolyzed to react with silica. Therefore, TESPT is very useful and nowadays well utilized in rubber industry [27, 40–43]. TESPT is added to the cross-linking reaction of SBR, followed by the sol-gel reaction of TEOS in the cross-linked SBR matrix to result in the cross-linked structure via in situ silica particles [42]. In the case of epoxidized rubber, on the other hand, amino group of silane coupling agent is reacted with epoxy group in situ during heat-pressing in the rubber production [43]. By this technique, very fine silica particles generated in situ are well dispersed. Furthermore, the use of silane coupling reagents becomes more and more important to modify and/or cross-link rubber molecules for affording the rubber network structure as well as the reinforcement effect.

2.2   Chemistry II: Cross-Linking Reaction

2.2.1   Invention and Development of Vulcanization

Almost all of rubber products have to be a three-dimensionally structured network for the sake of displaying a stable elasticity. In other words, they should be cross-linked in order to make them useful. In the field of rubber technology, vulcanization and cross-linking reactions have often been used interchangeably. Vulcanization was invented in 1839 [44, 45], but it was not recognized to be a series of chemical reactions until early in the twentieth century [46]. Cross-linking reaction is, therefore, a relatively modern word and is coexistent with vulcanization through the last century. In this book, however, vulcanization is to designate a kind of cross-linking reactions by which the cross-linking structure is composed of the sulfidic linkages, and the word "cross-linking reaction" is used as a general term including vulcanization. It includes, therefore, for example, the reaction using peroxides to result in the molecular network structure formation through the carbon-carbon linkage together with vulcanization, in which sulfur is used as the cross-linker. Historically, vulcanization using sulfur as a cross-linking agent is quite familiar, but a few reaction systems using sulfur donors such as tetramethylthiuram disulfide without elemental sulfur, which result in mono-sulfide linkage, are also included in the vulcanization.

In 1839, Goodyear invented the vulcanization, and then natural rubber (NR) became a useful material [47] He energetically repeated the experiments in order to make NR a material of use, in particular, to overcome the temperature changes: In winter, it turned rigid to loose soft and elastic characters, and on hot days in summer it was too soft. Goodyear did not know any chemical meanings of vulcanization: The summer problem is to be solved by network structure formation, which affords stable rubber elasticity. And, the winter one is due to low-temperature crystallization, which by chance is to be solved by vulcanization, too. Goodyear was lucky enough to invent vulcanization, which effectively and simultaneously resolved the two difficulties (see Sect. 3.5.2). After strenuous efforts, he finally invented the good combination of sulfur and white lead which enabled him to overcome the drawbacks of NR by vulcanization [44]. Remark 2 at the end of this chapter gives a brief account of his life.

Many kinds of metal reagents such as lead, magnesium, and zinc were used with cross-linking reagents including sulfur to accelerate the vulcanization reaction. The reason why he looked for the accelerator was the long reaction time of vulcanization when using only sulfur, ranging from several hours to a few days in order to get vulcanizates (see Fig. 2.6). The search of good accelerators has continued even up to now, in a sense. Among them all, zinc oxide remains selected nowadays not as an accelerator but as an activator in combination with stearic acid.

From the early twentieth century, the vulcanization has been regarded as a chemical reaction [46], and G. Oenslager (1873–1956), who recognized the chemical nature of vulcanization, found in 1906 that aniline accelerated the vulcanization more effectively than inorganic compounds [48]. His pioneering works were followed by aggressive trials of various organic amines (including even more or less acidic organic compounds in combination with organic bases) to get the best system for stably elastic rubber goods in a shorter time of vulcanization. Lots of these developmental efforts have led to sophistication of the design and technics of rubber vulcanization. Thiuram-type, guanidine-type, thiazole-type, and sulfenamide-type accelerators have been developed, and many of them are still utilized in rubber industry.

For examples, piperidinium pentamethylene dithiocarbamate was tested in 1912, zinc alkylxanthogenates in 1915, ZnMDC in 1918, TMTD in 1919, DPG in 1920, MBT in 1921, MBTS in 1923, MBS in 1932. In Fig. 2.2, such efforts to develop new accelerators are briefly explained [49]. At first, the accelerating influence of aniline is clearly shown when compared with sulfur alone. The vulcanization rate by using thiuram-type accelerator is the fastest among them, but it often is too fast to control the scorch during the compound processing. Sulfenamide-type accelerators are delayed-action accelerators, and their vulcanization rate is high. Note that their activity for vulcanization is not low at all, but they make the scorch time longer (delayed action). If necessary, use of PVI with accelerators enables us to secure the processing time easier. Due to the past efforts of rubber scientists and engineers, the vulcanization technique reached the stage of maturity early in the 1970s. Finally, the

Fig. 2.2 Comparison of cross-linking ability of various accelerators for sulfur vulcanization of natural rubber at 140 °C. The approximate age where the accelerators appeared in the market is shown within the parenthesis. Refer Fig. 2.3 for each chemical structure. From Fig. 9 of Ref. [49]

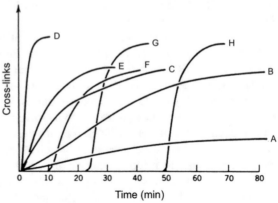

A: sulfur (1884)
B: aniline (1906)
C: 1,3-diphenyl-2-thiourea (CA) (1919)
D: tetramethylthiuram disulfide (TMTD) (1920)
E: 2-mercaptobenzothiazole (MBT) (1925)
F: 2,2'-dithiobenzothiazole (MBTS) (1925)
G: benzothiazolesulfenamides (1937)
H: benzothiazolesulfenamides + N-(cyclohexylthio)phthalimide
   (pre-vulcanization inhibitor (PVI)) (1968)

vulcanization systems using sulfur, organic accelerators [a typical one is the sulfe-namide-type accelerators such as N-cyclohexyl-2-benzothiazole sulfenamide (CBS)], ZnO, and stearic acid have established as a standard, particularly for tire manufacturing, by the end of the decade [50, 51].

T.S. Kuhn has chosen the term "paradigms" [52], who suggests that the term means "universally recognized scientific achievements that for a time provide model problems and solutions to a community of practitioners." He further has written, paradigms "suggest that some accepted examples of actual scientific practice—examples which include law, theory, application, and instrumentation together—provide models from which spring particular coherent traditions of sci-entific research." Additionally, he noted, "Acquisition of a paradigm and of the more esoteric type of research it permits is a sign of maturity in the development of any given scientific field." (Read Sect. 4.3.2, for more description on paradigm in relation with vulcanization.)

In accordance with the discussions by Kuhn, we may assume that the vulcan-ization paradigm was established late in the 1970s. After the technology has matured, rubber scientists and engineers have usually focused not much on developing new accelerators, but on the modification and/or improvement of the existing systems [53]. Based on the established vulcanization paradigm, it becomes possible now to select proper reagents for the vulcanization process. For example, efficient vulcanization (EV) and semi-EV has been much utilized in rubber industry, where the amount of sulfur was decreased compared to that of accelerators. In

Table 2.1 Recipe for CV, semi-EV, and EV systems of NR, and characteristics of vulcanizates

	CV	Semi-EV	EV
Sulfur (phr[a])	2.0–3.5	1.0–1.7	0.4–0.8
Accelerator (phr)	1.2–0.4	2.5–1.2	5.0–2.0
Accelerator/Sulfur	0.1–0.6	0.7–2.5	2.5–12
Polysulfidic and disulfidic cross-links (%)	95	50	20
Monosulfidic cross-links (%)	5	50	80
Cyclic sulfidic concentration	High	Medium	Low
Low-temperature crystallization resistance	High	Medium	Low
Heat-aging resistance	Low	Medium	High
Reversion resistance	Low	Medium	High
Compression set (22 h, at 70 °C) (%)	30	20	10

From Tables 4 and 5 in Ref. [54]
[a]Part per one hundred rubber by weight

Table 2.1, the characteristics of semi-EV and EV are summarized in comparison with the conventional vulcanization (CV) [54]. The first step of rubber processing is to choose the suitable vulcanization system among the three. However, it is notable that rubber experts have been dependent on the trial-and-error method in developing organic accelerators, little by logical designing on the base of clarified mechanism of the vulcanization reactions. A lot of chemical reaction mechanisms have been assumed on vulcanization, but there have not been any established mechanisms yet [49–51, 53–56]. The lack of information on the exact mechanism of vulcanization has been the reason why rubber engineers are still obliged to depend on the trial-and-error methodology in developing vulcanization accelerators.

2.2.2   Organic Accelerator System for Vulcanization

(1) Organic Accelerators

Rubber processing includes the vulcanization step, that is, it is a reactive processing to prepare a three-dimensional network structure for stable rubber elasticity. The exact shape and size of the final product are determined at the vulcanization step. Therefore, completion of the chemical reactions is shape-determining. Namely, the rubber compound (the mixture of rubber, vulcanization reagents, processing aids, and some others) has to flow at first to fill whole space of the mold, followed by a quick completion of the formation of three-dimensional network structure of rubber. Different from the processing of plastics, a chemical reaction plays the most important role in the processing of rubber. In other words, rubber processing is actually an example of reactive processing. The unique and the most difficult feature is the control of chemical reactions involved in vulcanization which are conducted in rubber matrix, not in a solvent as organic chemists usually choose.

Structure	Abbreviation	Name
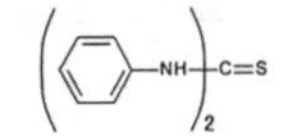	CA	1,3-diphenyl-2-thiourea N,N'-diphenylthiourea
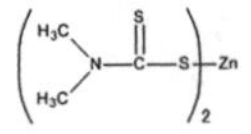	ZnMDC, ZDMC	zinc dimethyldithiocarbamate
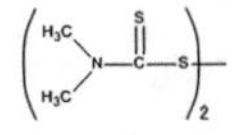	TMTD	tetramethylthiuram disulfide
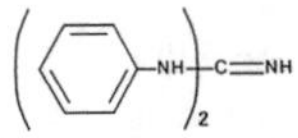	DPG	diphenyl guanidine
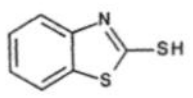	MBT	2-mercaptobenzothiazole
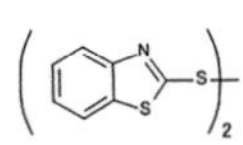	MBTS	2,2'-dithiobenzothiazole
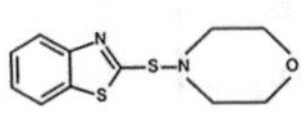	OBS, MBS	2-morpholinothiobenzothiazole
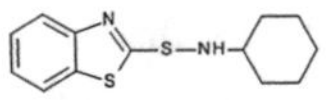	CBS	N-cyclohexylbenzothiazole-2-sulfenamide
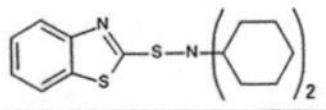	DHBS, DCBS	N-dicyclohexylbenzothiazole-2-sulfenamide

Fig. 2.3 Chemical structures, sample codes, and names of main accelerators

The chemical structures of representative accelerators are shown in Fig. 2.3. The main purposes in the development of accelerators have been the three features: (1) antiscorch property, (2) high reaction rate of cross-linking, and (3) low reversion tendency. Of course, the three features are requested to contribute to excellent mechanical and thermal properties of the resulted rubber vulcanizates. However, it often is too difficult to satisfy all these requirements to the same degree, and a kind of trade-off or compromise has to be considered for each rubber product. Practically, even the expert has been obliged to use the trial-and-error method, before adopting a definite strategy in designing the specific rubber processing. Surely, lots of the empirical knowledge is known as the rule of thumb.

For example, use of thiuram-type accelerators, the plateau region of which is very short, may afford very high reaction rate and good mechanical property. Thus, they are of use for vulcanizing saturated rubbers (relatively low unsaturation, compared with the diene rubbers) such as isobutylene-isoprene rubber (IIR) and ethylene-propylene-diene terpolymer rubber (EPDM). However, from a viewpoint of rubber processing, the scorch safety is regarded as the most important aspect during mixing, storing of rubber compounds, and shape forming by calender, extrusion, or compression molding. Hence, the sulfenamide-type accelerators such as CBS and *N,N*-dicyclohexyl-2-benzotiazolesulfenamide (DHBS) have been widely used from the early 1970s. Furthermore, antiscorching agents called as PVI have been also developed in order to improve the scorch safety during processing [51, 55].

From a viewpoint of sustainable development (SD), the spread of remaining vulcanization reagents and some of the by-products from them (e.g., by the wear of rubber tires) is to be noted more. Since the abrasive wear of rubber tires is a necessary function to minimize the wear of road surface, again a kind of trade-off is necessary. The global and integrated consideration is very important to decrease the environmental pollution on the Earth. Therefore, rubber scientists and engineers have to challenge for the further development of rubber material design, keeping SD in mind.

(2) **Activators for Vulcanization**

After the acceleration effect of aniline for vulcanization was found, the combination effect of metal oxide with sulfur/organic accelerator system was investigated, and zinc oxide (ZnO) was finally found as an excellent activator among the metal oxides, which was relatively easy to obtain. Once, zinc oxide had been used as a filling substance for rubber by many inventors including Goodyear in order to maintain physical stability of the rubber products and to give a white color, but it nowadays establishes its status as an excellent activator of vulcanization in combination with stearic acid. In this book, a term, "an accelerator activator," is not used, because the term is for the organic accelerator secondary to the main one in the vulcanization system. Zinc oxide is customary called zinc flower or zinc white, and is a white powder with a specific gravity of 5.47–5.78 g/cm^3. Ultrafine ZnO particles with smaller diameters and large surface areas are commercially supplied as active zinc oxide, and nowadays they are widely used together with stearic acid.

Recently, it was revealed that particulate ZnO is one of the main factors to control the network structure in the rubber vulcanizates, in particular, one of the origin of network inhomogeneity [57]. The new intermediate of zinc/stearate complex, which was generated in situ by the reaction of ZnO and stearic acid in the rubber matrix [58], was to form the mesh network of the rubber matrix in the vulcanizate. In addition, the ZnO cluster was found to form the network domain of high network-chain density by absorbing sulfur and accelerator on the surface of particulate zinc oxide. Thus, the inhomogeneous two-phase network structure was the result of the vulcanization reaction using zinc oxide. Read Sect. 4.3.1 and Ref. [57] for more details.

Stearic acid is a linear saturated hydrocarbon whose number of carbon is 18. Therefore, it has been assumed that the acceleration effect of stearic acid to the vulcanization reaction is due to the increasing solubility of zinc ion into the hydrophobic rubber matrix after generating zinc stearate. Though this assumption may still be playing a role, the more important and acceptable concept is as follows:

The structure of reactive intermediate, which is generated after the dissolution, is "the complex of the ratio of zinc/stearate = 1/2," whose zinc ion is coordinating with sulfur and anion fragments from the reacted accelerator [58]. This intermediating complex is supposedly playing the most important role in promotion of the sulfur cross-linking reaction, i.e., the generation of –C–S– linkages.

The final step of vulcanization is the elimination reaction of zinc ion as zinc sulfide (ZnS) [50, 51, 56, 58]. Since zinc sulfide was experimentally detected, the vulcanization reaction scheme involving zinc has widely been accepted by rubber scientists. However, it has been very difficult to specify the structure of the reactive intermediates because the reaction medium (solvent) was a high-viscous rubber matrix composed of high molar mass rubber chains. This is one reason why zinc oxide and stearic acid have been named as an accelerator "activator" for a long time, although they are indispensable reactants in vulcanization. For more details of recently proposed new intermediate, refer to Sect. 4.3.1 and Ref. [58].

It is notable that the mineable years of zinc oxide as resource are currently estimated to be about 20 years. Therefore, the search of an alternative metal ion has to be promoted, which is surely one of most important challenges for the future of rubber industry, hence for our future society in general where the transportation is more and more essential from SD viewpoint.

2.2.3 Cross-Linking Reactions by Peroxides and Others

(1) Peroxide Cross-Linking

The application of vulcanization (cross-linking using sulfur) is limited to diene rubbers or at least those having some carbon-carbon double bonds in their rubber chains or as pendant groups. Among non-sulfur curing systems, organic peroxides are now commonly used to cross-link polyethylene, EPM or EPR, and many of polyolefinic elastomers. As noted briefly in Sect. 2.1, novel polyolefin rubbers have recently been significantly developed by hydrogenation of conventional diene rubbers and by polymerizations using metallocene-typed catalysts, and so on. It means the demand for highly heat resistant rubbers is more and more increasing. Hence, the application of peroxide cross-linking to the processing of them is expanding. The peroxide cross-linking is applicable not only to saturated rubbers but also to unsaturated rubbers [59, 60]. In the reaction mechanism of peroxide cross-linking, free radicals generated by the homolysis of peroxide extract hydrogen atoms at the allyl position and/or tertiary hydrogen atoms in rubber molecules, and followed by radical coupling and/or addition reaction to form a cross-linked structure (C–C linkages). The cross-linking reaction by coupling often occurs in isoprene rubber (IR), and the chain addition reaction occurs in butadiene rubber (BR) and so on. In order to increase the cross-linking efficiency by prolonging the lifetime of the generated free radicals, cross-linking coagents are commonly used

[61, 62]. There appeared a paper reporting that the deproteinizing effect promotes radical diffusion in the peroxide cross-linked NR latex and gives more uniform network structures [63]. In addition, peroxide cross-linking is simpler in formulation and processing steps and the number of added reagents and steps are smaller than the sulfur cross-linking reaction (vulcanization). For example, in a study on ionic conductive elastomer materials made from high molecular weight branched poly (oxyethylene), peroxide cross-linking using benzoyl-*m*-toluoyl peroxide and *N,N-m*-phenylene bismaleimide brought to a superior polymer solid electrolyte [64]. (See Sect. 3.3.3.) In some industrial applications, both sulfur cross-linking (i.e., vulcanization) and peroxide cross-linking are used in combination. This may be an example of one step further development from the current vulcanization paradigm.

(2) **Cross-Linking Reactions of Non-peroxide Type**

The peroxide cross-linking is also used for saturated rubbers such as silicone rubbers, fluoro rubbers, and so on. On the other hand, when the peroxide cross-linking is applied to polyisobutylene (PIB) and butyl rubber (IIR), the main chain scission (not at random, but the zipper-type scission) occurs frequently. Therefore, IIR is subjected to not only sulfur cross-linking but also quinoid and resin cross-linkings. Thiourea curing is a general cross-linking method for chlorine-containing rubbers such as chloroprene rubber (CR) and epichlorohydrin rubbers (CO, ECO), and it is considered that bis-alkylation reaction would occur under the presence of acid accepter. Due to the toxic problem of ethylenethiourea (EU), polyol curing is now applied to these rubbers. Furthermore, polyamine and polyol cross-linking, in which the acid accepters are metal oxide or metal hydroxide, are used for chlorinated rubbers and fluoro rubbers. The cross-linking reaction of silicone rubber is achieved either by using peroxide or by the hydrosilylation reaction by using a platinum catalyst under the presence of di-functional low molar mass oligomer which carries a vinyl group at both chain ends.

For conventional rubbers, the cross-linked polymers can be obtained by sol-gel reaction under water after modification of introducing the alkoxysilyl groups using silane coupling agent [65]. This is sometimes called moisture cure or silane cross-linking. Among silane coupling agents, bis [3-(triethoxysilyl) propyl] tetra-sulfide (TESPT) is an industrially valuable reagent because it functions as both a cross-linking reagent and a coupling reagent [27, 30]. Di-functional reactive polymers, which carry a reactive group at the both chain ends (telechelic), are reactive toward reagents whose number of functional end group is three or more to generate the three-dimensional network structure. For example, telechelic olefin-based liquid rubbers or diene-based liquid rubbers carrying hydroxyl groups are used for preparing cross-linked products by urethane linkages. By using these reactive di-functional prepolymers, it is possible to produce a network structure without any free terminal chains, i.e., dangling chains that do not contribute to rubber elasticity. This technique is considered to be a good method for preparing the cross-linked samples with a model cross-linked structure, and thus, it is often used for academic discussing on rubber elasticity [66–69].

2.3 Physics: Rubber State and Rubber Elasticity

2.3.1 Rubber State

The thermodynamic state of polymeric substances in general is shown in Fig. 2.4, in which the volume (V) is plotted against absolute temperature (T). The slope at the central region in the figure is the volume expansion coefficient of crystalline solid state. When the increasing T reaches the melting point (T_m), we observe a sudden jump of V which is a noncontinuous change of V. This is the solid-to-liquid transition, and further increase of T gives rise to further increase of V with a larger slope (the volume expansion coefficient of liquid) than that of solid states. Inversely, when a substance at liquid state is subject to cooling, V suddenly drops at T_m non-continuously to crystalline solid state (T_s, solidification temperature), to continue volume decrease. That is to say, the transition is reversible. If the temperature change is quasi-statically conducted, the relationship, $T_m = T_s$ holds.

Among majority of polymers, rubbers in particular, are not much crystalline but fundamentally amorphous, and the cooling of polymer melt (at an amorphous liquid state) results in the disregard of T_s, to continue the decrease of V down to glass transition temperature T_g (see Sect. 1.3.2 for T_g). At T_g, there observed noncontinuous sudden change of not V itself but of the slope (mathematically, a derivative) of volume change. The state below T_g is called glassy state and the above is rubbery state. This transition might have been called the glass-rubber transition [70], but

Fig. 2.4 Two-dimensional volume (V) versus temperature (T) plot of polymeric materials to show their thermodynamic states. (Prepared by S. Kohjiya)

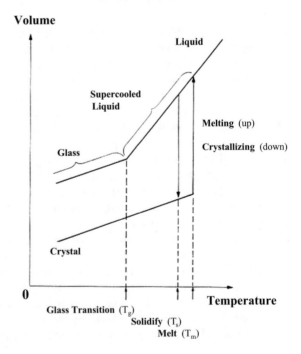

"the glass transition" is still widely used, because glass has historically been known for much longer time than rubber. The T_g is the lowest limit for the utilization of rubbers, but it is the highest one for thermoplastics, i.e., T_g is approximated to be a softening temperature of the plastics. In other words, at the temperature above T_g, plastic materials are too soft for utilization, while soft rubbery state above T_g is the utilization temperature range for rubbers. Many of the plastics are semi-crystalline, and both T_g and T_m (or T_s) may be observed. Also, a few polymers have been crystallized under specific conditions to give a polymer single crystal [71] (see Sect. 3.4).

The glass transition is a kind of solid-liquid transition, but it does not show noncontinuous volume change: Only the slope (i.e., differential of V by T) of V versus T curve does. Based on this fact, it is often called the second-order thermodynamic transition. However, experimental observation of T_g is much dependent on the thermal history of the sample, and the transition is not considered exactly thermodynamic. Therefore, a simple description of the glass transition in accordance with the free volume theory [70, 72, 73] and the WLF (Williams, Landel & Ferry) equation [74, 75] is given here.

The specific volume (volume/weight) V of a matter is assumed to be the sum of occupied volume V_0 and free volume V_f. V_0 is based on the volume by the van der Waals radii of the component atoms inclusive of the volume allowance due to vibration, and the residual volume is named as a free volume. That is,

$$V = V_0 + V_f \qquad (2.1)$$

Free volume has experimentally been quantified by the positron annihilation lifetime spectroscopy (PALS). The presence of positron, antiparticle of electron, was theoretically predicted by the Dirac's equation, which annihilates with electron to irradiate photon. Since positron is positively charged and repelled by proton, it is annihilated by an electron in the outer shell of an atom. On these basics, lifetime of positron in an insulating material is assumed to depend on the presence of free volume: The more is the free volume, the longer is the lifetime of positron. PALS data [76] suggest that upon heating above the T_g, the number of holes does not increase, but their volume does.

The PALS results suggest that V increases by heating from absolute temperature zero, and the slope of increase is enhanced at a certain temperature T_r. This acceleration is explained by that of free volume: Fraction of free volume f is expressed as $f = V_f/V$, and the next equation may explain the acceleration,

$$f = f_r + \alpha_f(T - T_r) \qquad (2.2)$$

where T is absolute temperature, T_r is a reference temperature, f_r is the fraction f at T_r, and α_f is the expansion coefficient of free volume at the higher temperature than T_r which is assumed to be much larger than at the lower temperature. Taking this behavior of free volume into account, Doolittle proposed that zero shear viscosity of liquid is expressed as

$$\ln \eta_0 = \ln A + B \frac{V - V_f}{V_f} \qquad (2.3)$$

where A and B are constants.

On the other hand, Tobolsky et al. [77] have reported the famous results shown in Fig. 2.5, which shows relaxation modulus $E(t)$ measured in the temperature range between −80.8 and +50 °C. The observed curves were shifted using −65.4 °C as the reference temperature; those at higher than −65.4 °C were shifted to right-hand side and those at lower to left-hand side to result in the broken line, which is called the master curve. The displacement is $\log \alpha_T$, and α_T is called the shift factor. It depends on temperature as shown in the upper-right inset. Naturally, $\log \alpha_T$ is zero at −65.4 ° C. Figure 2.5 suggests a very interesting result, that is, the obtained master curve has managed to cover the range of 10^{16} s, while the experimental one covered only 10^5 s. Considering that our lifetime might cover only 10^9 s at most, without the shifting we would not have managed to get the results in Fig. 2.5. The principle used here is called the time-temperature reducibility.

William, Landel, and Ferry reported that the behavior of α_T of many polymers was well described by the following equation [74], now called the WLF equation [72, 73]:

$$\log \alpha_T = \log \frac{\eta_0}{\eta_{0r}} \left(\frac{T_r \rho_r}{T \rho} \right) = -\frac{C_1 (T - T_r)}{C_2 + (T - T_r)} \qquad (2.4)$$

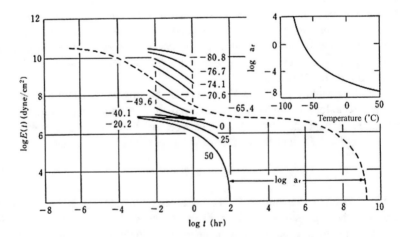

Fig. 2.5 Preparation of master curve by use of time-temperature reducibility from time (frequency) dependence of relaxation modulus $E(t)$ of polyisobutylene, and temperature dependence of shift factor α_T shown in the inset. (From Fig. 11 in Ref. [77])

where the subscript r means the value at the reference temperature T_r and C_1 and C_2 are constants. The ratio $T_r\,\rho_r/T\rho$ (here, ρ is density) is practically equal to unity. They have chosen $T_r = T_g + 50$, and reported the values, $C_1 = 8.86$ and $C_2 = 101.6$. When we use T_g for T_r, we obtain

$$C_1 = 17.44, \quad C_2 = 51.6 \tag{2.5}$$

From Eqs. (2.3) and (2.4), the following equation is derived

$$\log \alpha_{\mathrm{T}} = \frac{B}{2.303}\left(\frac{1}{f} - \frac{1}{f_r}\right) \tag{2.6}$$

By comparing Eq. (2.6) with the WLF equation (2.4), the following relations

$$C_1 = B/(2.303\,f_r), \quad C_2 = f_r/\alpha_f, \quad C_1C_2 = B/(2.303\,\alpha_f) \tag{2.7}$$

are resulted. Using the values in Eq. (2.5), the following two values are obtained.

$$f_g = 0.025, \quad \alpha_f = 4.8 \times 10^{-4}(\deg^{-1}) \tag{2.8}$$

That is, at the glass transition temperature, free volume may occupy 2.5% of the total volume. In other words, T_g is considered to be a temperature of iso-free volume state of polymeric materials. The discussion given above is historically the first approximation, and the glass transition of polymer is recently treated as a local relaxation process [73].

Return to Fig. 2.5, and consider the state of materials at 0 K and onward. With the increase of temperature, the volume is expanding and intramolecular movements (rotation and vibration) may be activated to reach T_g. The free volume is about 2.5% of the total volume at T_g, and micro-Brownian motion (segmental Brown movement; a segment may be composed of several consecutive monomeric units in the rubbery chain) becomes active. With temperature, this thermal agitation of the segmental movement is activated to result in a high-viscous liquid state. As shown in the figure, supercooled liquid state and that at higher temperature are in rubbery state. The viscosity is lowered by increasing the temperature, and still the rubbery states look apparently like a solid due to its highly viscous nature. In other words, the rubber state is microscopically liquid, but it is macroscopically solid. The phenomenon called "cold flow", which is observed on a solid amorphous rubber (e.g., SBR) when it is kept standing for a year or so, would display its liquid nature. Below T_g, the segmfental micro-Brownian motion is frozen, and the rubber is an amorphous solid of a glassy state.

Rubber state is, therefore, not a state specific to rubbers, but almost all polymers of high molar mass may be rubbery above T_g [72, 73, 78], or above T_m for crystalline polymers. As explained in the next subsection, the most impressive feature of the rubber state is its mechanical behavior, i.e., rubber elasticity. An important scientific conclusion here is both rubber state and rubber elasticity, which are characteristics of polymers in general above T_g, and it is to be remembered that the two are not in particular for rubbers. The rubber simply shows them at room temperature.

2.3.2   Rubber Elasticity (Entropic Elasticity)

At the temperatures above T_g, a polymeric chain is in a random conformation (as a whole it is a random coil) as schematically shown in Fig. 2.6, and each segment consisting the chain is in an active motion due to micro-Brownian movement, which increases parallel with the increase of temperature. In the figure, one end of the chain A is fixed at the origin of the coordinate axes, and the other end is situated at point B (x, y, z) or (R, θ, φ) in the polar coordinate. The chain is in a randomly dynamic state, and the probability of the end situated at point B may be calculated. This probability $p (x, y, z)$ is defined as the possibility to be found in a small volume, $dv (=dxdydz)$ situated at point B, which is

$$p(x, y, z)\, dxdydz = \left(\frac{b}{\sqrt{\pi}}\right)^3 e^{-b^2 (x^2 + y^2 + z^2)} dxdydz \tag{2.9}$$

For the chemical structure of the chain, the model shown in Fig. 2.7 is adapted to elucidate constant b in Eq. (2.9) as

$$b^2 = 3/(2Nl^2) \tag{2.10}$$

where N is degree of polymerization and l is the bond length of C–C. Equation (2.9) is the Gaussian function, and the chains whose conformation is represented by this equation are the Gaussian chains, or the random chains since their motion is not regular but random. The end-to-end distance of this type of chains, i.e., the probability of distance AB be between R and $R + dR$ is

Fig. 2.6 Sketch showing random coil conformation of a polymeric chain. Modified Fig. 3.4 in Ref. [79]

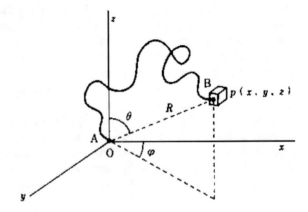

Fig. 2.7 A conformational molecular model of monomeric unit in the polymer chain. (Modified Fig. 74 in Ref. [80])

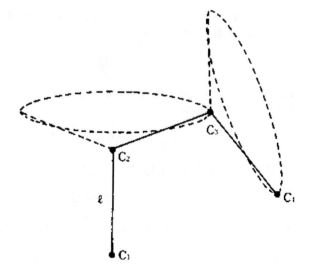

$$P(R)dR = \frac{4b^3}{\sqrt{\pi}}R^2 e^{-b^2 R^2} dR \tag{2.11}$$

And the root-mean-square of R is

$$\langle R^2 \rangle = \frac{3}{2b^2} = Nl^2 \tag{2.12}$$

Taking root of the both sides of Eq. (2.12) gives

$$\sqrt{\langle R^2 \rangle} = \sqrt{N} \times l \tag{2.13}$$

According to Eq. (2.13), the end-to-end distance of the Gaussian chain is much smaller than that of fully extended chain ($=N \times l$). (Here, note that R is the distance or length, not the gas constant.) N is equivalent to the degree of polymerization, and if $N = 100$, Eq. (2.13) gives $10 \times l$ which is one-tenth of $100 \times l$. When we assume the points A and B in Fig. 2.6 are the cross-linking points, the Gaussian chain AB is a network chain. In this case, the cross-linked rubber (=elastomer) may be stretchable to the ten times length of the original. Few solid materials other than rubbers or elastomers are stretchable to 10% (to the original length $\times$ 1.1). The reversible high extensibility of rubber is standing alone among so many solid materials.

Thermodynamic reasoning of rubber elasticity affords another reason why the rubber is able to display so unique a mechanical behavior. When a cross-linked rubber of length R and volume V is elongated, the rubber sample is assumed to absorb ΔQ of heat and its internal energy is assumed to be increased by ΔE. The first law of thermodynamics (law of conservation of energy) suggests that the following relation holds

$$\Delta E = \Delta Q + f\,\Delta R + P\,\Delta V \tag{2.14}$$

and ΔV may be negligible since the Poisson ratio of rubber samples is 0.5, which means $\Delta V = 0$. Hence,

$$\Delta E = \Delta Q + f\,\Delta R \tag{2.15}$$

The change of entropy by the elongation is

$$\Delta S = \Delta Q / T \tag{2.16}$$

per the second law of thermodynamics. From Eqs. (2.15) and (2.16),

$$\Delta E = T\,\Delta S + f\,\Delta R \tag{2.17}$$

is derived.

In the deformation of cross-linked rubbers, the term $P\,\Delta V$ is negligible, and further discussion is on Helmholtz free energy, which is defined as

$$F = E - TS \tag{2.18}$$

From Eqs. (2.17) and (2.18),

$$\Delta F = \Delta E - T\,\Delta S = f\,\Delta R \tag{2.19}$$

That is to say, the following relation is derived.

$$f = \left(\frac{\partial F}{\partial R}\right)_{T,V} = \left(\frac{\partial E}{\partial R}\right)_{T,V} - T\left(\frac{\partial S}{\partial R}\right)_{T,V} \tag{2.20}$$

where the symbol Δ, a possible smallest amount, is converted to a partial differential symbol ∂. Equation (2.20) suggests that under the conditions of constant T and V, the force f which is inducing the deformation of the rubber sample consists of two terms: One is the change of internal energy $(\partial E/\partial R)_{T,V}$ and the other is that of entropy $(\partial S/\partial R)_{T,V}$. Meyer and Ferri [81] reported a classical experimental results in 1935, which suggested the term $(\partial E/\partial R)_{T,V}$ in Eq. (2.20) was negligibly small above 230 K (T_g of the rubber sample they used) to result in the following relationship,

$$f = \text{constant} \times T \tag{2.21}$$

In other words, rubber elasticity is entropy controlled, but is not energy controlled, which is firmly elucidated both theoretically and experimentally. Internal energy has been the governing factor of elastic deformation in predominant materials used so far.

Entropy is defined by the Boltzmann equation in statistical thermodynamics (or statistical mechanics) as follows [82–84]:

$$S = k \log \Omega \tag{2.22}$$

where k is the Boltzmann constant, log is natural logarithm, and Ω is the number of possible conformation of the rubber chain AB shown in Fig. 2.6. When the chain AB is fully extended (the length is $N \times l$) by applying a force f, AB would be linear and the possible conformation is only one. In other words, $\Omega = 1$, hence $S = 0$. When the distance AB is smaller than $N \times l$, the possible conformations (i.e., possible routes connecting A and B) increase, and the entropy S increases, too. The entropy of the Gaussian chain in Fig. 2.6 is decreasing with its elongation, and at the fully stretched state $S = 0$. If the force f (stress) on the chain be removed, the chain would return to the original high entropy state in accordance with the second law of thermodynamics. Elongation of the rubber is a process against the second law. Consequently, we have to do a thermodynamic work W (i.e., $\Delta W = f \times \Delta l$) in order to elongate a rubber sample.

Taking Eq. (2.21) into account, we get the following equation from Eqs. (2.20) and (2.22).

$$f = -T k \log\left(\Omega_f / \Omega_i\right) \tag{2.23}$$

where Ω_i and Ω_f are the numbers of possible conformation at the initial state and the stretched state, respectively. Ω is a function of the end-to-end distance R of the chain, and the entropy S is expressed from Eqs. (2.20) to (2.21)

$$S = \text{constant} - kb^2R^2 \tag{2.24}$$

Accordingly, Eq. (2.23) is converted to

$$f = 2kTb^2R \tag{2.25}$$

This relationship suggests that f is proportional firstly to R and secondly to T. The second is worth mentioning: The proportionality to T is against the elasticity of usual solid body in general. Their elasticity is of energy origin, and f (at a certain strain) decreases with the increase of temperature. The entropic elasticity force of the Gaussian polymer chains is proportional to T, entirely different from the energy-controlled elasticity. This unique behavior of rubber elasticity is only shared with ideal gas, not with any solid materials so far.

Rubber elasticity has been explained on one rubbery chain or on one network chain in the cross-linked rubber. However, our measurements of rubber elasticity are usually on cross-linked rubber samples, not on a single chain. On the treatments of rubbery network systems, study more advanced references [79, 80, 85–87] in order to reinforce your ability for understanding rubber elasticity more, which is a must for your study on rubber science. One of the most fundamental results from the classical rubber elasticity theory is Eq. (2.26), i.e., the stress σ for stretching to elongation ratio α ($\alpha = l/l_0$, l_0 is the initial length and l is that after stretching) is given by

$$\sigma = vkT(\alpha - \alpha^{-2}) \tag{2.26}$$

where v is the network-chain density. Here, surprisingly only one parameter v is enough to know on the rubber sample in describing the tensile mechanical behaviors of rubber. In other words, the basic tensile behavior of almost all commercially available rubbers listed in the table in "Appendix", has been described by using v regardless of their difference in chemical structures. Equation (2.26) is, in spite of its simplicity, versatile enough to approximately describe tensile behaviors of rubbery materials in general, and hence it seems to be based on an essential principle of nature from the thermodynamic viewpoint.

In this case, the deformation is assumed to be affine and rubber is non-compressive, i.e., size of the rubber sample under stretching is (αx_0, $y_0/\sqrt{\alpha}$, $z_0/\sqrt{\alpha}$), hence no change of volume by deformation. Classical theory of rubber elasticity including Eq. (2.26) was proposed by Meyer, Kuhn, Guth, Wall, James, Treloar, and Flory [79, 80, 85–88]. [For more details, see Sect. 2.3.4(3).] These and some other studies on rubber elasticity had given rise to the establishment of the classical theory around 1940. Thus, assuming the Gaussian chain as an idealized rubber chain, thermodynamic and statistical thermodynamic considerations have given rise to the classical theory, which elucidates the following unique features of rubber elasticity:

(i)   Rubber is deformed by a small stress to result in a very low Young's modulus, possibly due to entropy effect, compared with the other materials whose elasticity is due to energy effect.

(ii) Extensibility up to the rupture is extremely high, which is based on the dynamic state of rubber (under micro-Brownian segmental motion), described by Eq. (2.23). (Also see **c**. High extensible ability may be related with the incompressibility.)

(iii) Volume change associated with deformation is very small, i.e., the Poisson's ratio is 0.5. The ratio is 0.2–0.25 for ceramics, ca. 0.3 for metals, and about 0.4 for woods and plastics. No volume change even at large strain up to 10 times elongation of the original, is really surprising, the reason of which is that rubber is in a dynamic state by the micro-Brownian motion to be a polymeric solvent. This factor also gives rise to the toughness of rubber, since cavitation (one of the reasons of fracture of plastics under strain) is avoided by liquid-like nature.

(iv) As shown in Eq. (2.25), the stress at a certain elongation increases with absolute temperature. That is, it hardens by heating, which is unique and not observed in other materials at all. This feature is especially due to the origin of rubber elasticity, the entropic change.

(v)   The last characteristic is the reversibility at elongation-contraction cycles, which is ascribable to the network structure of cross-linked rubber as well as the dynamic state even at a high strain just below the mechanical rupture.

These features of rubbers are all in their own right, and ascribed to the entropic nature of deformation of rubber. Rubber elasticity is the only one example of entropy-driven elasticity at solid state, and the elasticity of gas (pressure of the ideal gas) is the other example. The unique and high performance device in transportation, a pneumatic tire for automobiles and aircrafts, has its theoretical base for its unique function, that is, the tires are a skillful combination product of rubber and air, the elasticity of which is of entropic origin.

2.3.3   Unique Role of Rubber in Elucidating the Nature of Macromolecule

The essential prerequisite for establishing macromolecular science as a branch of science was, of course, the elucidation of high polymeric nature of macromolecules, experimentally and theoretically. This requirement is described here, followed by more or less theoretical and historical aspects on rubber elasticity in the next Sect. 2.3.4. Both were destined to conclude the establishment of macromolecular science early in the 1940s.

Historical complications surrounding the birth of polymer science are briefly described by Flory in the first chapter of his book entitled "Principles of Polymer Chemistry" [80] and much more in detail by Furukawa [88]. Since the last decade of the nineteenth century, colloid had been understood as a new intermediate phase between the microscopic dimension and the macroscopic bodies. It was also assumed that the colloidal state (an aggregated state due to a certain secondary valence, not due to covalent bonding) might be realized by any substances, at least theoretically. Around the beginning of the twentieth century, the idea of colloidal states due to the secondary valence force was so influential among physical and chemical scientists that the acknowledgment of macromolecules as the substance of high molar mass not by colloidal association but by the main valence (i.e., covalent bonding) was delayed to the 1930s. In other words, a macromolecule can show a colloidal behavior even without association did not enjoy a widespread acceptance until then.

Under these situations, it has not been much noted that rubber had played an important role toward the establishment of the molecular concept on macromolecules or high polymers (for this concept, "macromolecular theory" [89] or macromolecularity is used here). Early in the twentieth century, spread of automobiles stimulated the demand for rubber tires, hence natural rubber (NR). NR is an agricultural product of tropical countries, and its marketing was under the British control. Consequently, the development of synthetic rubbers was focused on chemistry arena. A German chemical engineer, F. Hofmann (1866–1956) at Bayer, submitted a plan for synthetic rubber to the directors' board, which was accepted. His group synthesized dimethylcyclooctadiene (DMCOD, a cyclic dimer of isoprene), which had been proposed to be the chemical structure of NR by Prof. C.D. Harries (1866–1923). Harries had estimated NR was a colloidal aggregate of DMCOD from the results of ozonolysis (oxidative degradation reaction by ozone) of NR. There appeared a counterargument by S. Pickles (1876–1962), but it was not persuasive enough to cancel the colloid paradigm, i.e., colloids as an intermediating new dimension between macro- and microsubstances. The colloid paradigm was influential and accepted widely then. However (or naturally, we can say now), Hofmann failed to get NR from DMCOD. If his experimental details of the results were published in a scientific journal, Harries' colloidal NR might have been negated much earlier.

A German organic chemist, H. Staudinger (1881–1965), was stimulated by these arguments on NR, and carefully studied them. He probably believed in the covalent bonding to form macromolecules, and began new project on them from the viewpoint of chemical reaction. In 1920, he published a paper entitled "On the polymerization," followed by a paper "Hydrogenation of rubber and its structure" in 1922, in which he proved no influence of hydrogenation of double bonds in NR on the macromolecularity. He conducted vigorous and tireless research on macromolecular nature of various polymeric compounds, to establish the macromolecular theory early in the 1930s [90]. He enthusiastically continued his organic chemistry approach to macromolecules with his students and researchers. Many engineers of German chemical companies were also cooperated with him. Even though he did not have a friendly relationship with the Nazis at all, he was supported by some chemical companies, and he managed to continue his active research even through the war time. His great effort gave rise to winning of the Nobel Prize in Chemistry for his discoveries in the field of macromolecular chemistry in 1953 [89, 90].

In addition to the tireless studies by Staudinger and his coworkers in Germany, a series of elegant papers on polymer synthesis by W.H. Carothers (1896–1937) at Du Pont Company contributed to the final establishment of the macromolecular theory during 1930s, particularly in the USA. Carothers was the inventor of chloroprene rubber (CR) and nylon-6, 6. The former is the first commercial synthetic rubber, and the latter is one of the first fully synthetic fiber materials. On the developmental chemical works of these polymers, he was so talented to do excellent scientific studies and to publish them, just in order to overcome the stumbling stones toward the final developmental objectives. His papers on polymerization have built the fundamentals for the synthetic chemical aspects of polymer science [91], together with those by Staudinger. Incidentally, P.J. Flory (1910–1985), a physical chemist, was once in Carother's group at Du Pont. Flory and lots of other scientists in USA are said to have accepted the macromolecular theory not by Staudinger's but by reading Carothers' papers [88, 91]. Chapters on synthetic polymer chemistry in Flory's famous book [80] are mainly based on Carothers' publications [91], which he studied at Du Pont, not much based on Staudinger's [89, 90].

2.3.4   Contribution of Rubber Elasticity Theory to Establishing Macromolecular Science

(1) X-ray Diffraction Studies of NR in the 1920s

Surprisingly, the earliest origin of contribution of rubber elasticity to the macro-molecularity was a structural study of NR by X-ray diffraction technique. Father and son Braggs elucidated the Bragg's condition for utilizing wide-angle X-ray diffraction (WAXD) data to calculate the crystal structure in 1912, followed by lots of structure determinations using this technique (see Sect. 3.4). As early as in 1925, J.R. Katz (1880–1938) published results of WAXD measurements on NR samples under stretching [92, 93]. (Possibly, this might be the first trial toward dynamic WAXD measurement.) He ascertained that the observed diffraction patterns were not due to any impurities and recognized the orientation of rubber chains, which suggests that he was already at the side of Staudinger. About that time, M. Polanyi (1891–1976) and H. Mark (1895–1992) started WAXD studies on rubber, cellu-lose, and others. Mark moved to I.G. Farbenindusries, and restarted the X-ray study with K.O.H. Meyer (1883–1952) to determine the crystal structures of NR and gutta-percha [94]. They recognized that lattice constants of the unit cells of NR and gutta-percha were consistent only with *cis*-1,4-polyisoprene and *trans*-1,4-polyisoprene, respectively. The two are among the earliest results on the polymer crystal models. Initially, the crystal lattice was supposedly aggregated (not by covalent bonding) to form the polymer, but Mark eventually admitted these X-ray results were fully compatible with the macromolecular theory, and sat at Stauginger's side around 1930.

(2) Concept of Flexible Polymer Chains

The other preliminary background of rubber elasticity theory was the concept of flexible polymeric chains, followed by the thermodynamic and statistical treatments on rubber elasticity. Rubber elasticity, which was a unique physical property of polymers, was much highlighted in the 1930s and 1940s. Staudinger, who was an organic chemist, was interested particularly in viscosity among so many properties of polymers, believing that the polymeric chains were rod-like: He had been sug-gesting the linear relationship between relative viscosity (η) and molar mass (M) of the polymer, i.e., $\eta/c \sim M$ (c is the concentration of polymer). This relation was later accommodated in the Houwink-Mark-Sakurada equation,

$$[\eta] = K M^a \tag{2.27}$$

where $[\eta]$ is intrinsic viscosity, K is the constant specific to each polymer-solvent pair, and the exponent a is the constant, too. (Numerically, a is between 0.5 and 1.0.) The Staudinger's equation is only for a special case, $a = 1$, which holds in the case of rigid rod-like polymers, not applicable to flexible chains like rubbers where a is much smaller than unity down to 0.5. Furukawa noted [88] that Staudinger showed wooden sticks as a model of his macromolecule in his lecture for many years. At the Faraday Society Discussion Meeting in 1932, a coming out party for Mark in England by chance, Mark lectured validity of the macromolecular theory of Staudinger [95]. Some polymer scientists have regarded that macromolecular chemistry, as a new branch of science, is publicly recognized at this meeting. However, this recognition was regrettably smeared by the right person himself, i.e., by the repeated Staudinger's claim of the rigid rod concept of macromolecules there. Mark opposed this claim from physicochemical viewpoint, of course. All through these discussions, relevant physical scientists got interested in the flexibility of polymer chains. Thus, came onstage a question, "How can we explain rubber elasticity by the concept of flexible polymer chains?"

(3) **Classical Theory of Rubber Elasticity and Establishment of Polymer Science**

Here, a few German, American and English groups (Guth moved from Vienna to the United States in 1937) appeared on the stage to report on thermodynamically and statistically elaborated rubber elasticity theory; Meyer et al. [96], Kuhn [97–99], Guth and Mark [100], Wall [101], Guth and James [102], Treloar [103], James and Guth [104], Flory [105], and others. Among those papers, the 1932 Meyer-Susich-Valko paper [96] was the first. The 1934 Guth-Mark paper [100] was most favorably evaluated by A.J. Staverman in Leiden as

"One of the great strides forward in the history of science, comparable to the conception of the kinetic theory of gases"

which is quoted in Ref. [88] (p. 260), while later majority of scientists tended to prefer Kuhn's treatments [97–99]. For example, G.V. Schultz (a physical chemist in Staudinger's group) and Flory (a physical chemist in Carothers' group) were both impressed by Kuhn's papers, and both were encouraged to apply mathematics to polymers by their bosses as vividly described in Ref. [88] by Y. Furukawa. Separately and in isolation during the World War II in Tokyo, a young physicist R. Kubo, University of Tokyo, suggested a few mistakes in Kuhn's publication and he developed his unique statistical thermodynamic treatments of rubber elasticity by himself [106]. His original might be a trial of renewing for the betterment of Kuhn's treatment. Also, Flory had elaborated in extending Kuhn's work to the point of perfection, of course, without knowing Kubo's effort. These were only a part of exciting scenes now in the history, when a new scientific discipline was giving a birth.

Consequently, the quoted unstinted praise to the classical rubber elasticity theory by Staverman is to be accepted fully, as a eulogy toward all the publications [96–106] mentioned above. Anyway, the classical theory of rubber elasticity based on the statistical thermodynamics was firmly established during the 1940s by the spread of these publications. This achievement has given the final and the most important touch to a new discipline, to result in the birth of "Macromolecular Science" as a new face among modern science disciplines. It is notable that Staudinger had clung to "Die Makromolekulare Chemie," and Makromolekül (or macromolecule in English) is preferred to polymer in Germany still now, while "polymer science" and "polymer chemistry" are popular in the USA. The readers are urged to study the classical rubber elasticity theory at least by one of the textbooks of polymer science [79, 80, 107], to be followed by an advanced text-books [108] or by studying further development by one of the specialized mono-graphs [73, 86, 87, 109, 110].

(4) **Prospect of Molecular Theory of Rubber Elasticity**

Further progress of the theory of rubber elasticity is to be discussed a little, now, more than 60 years after Flory's book was published [80]. The classical theory was surely a molecular theory based on a much simplified Gaussian (or random flight) model, and successful in describing the unique features of rubber elasticity in spite of its simplicity. However, the progress in the arena of rubber elasticity was directed to the so-called phenomenological approach [79, 111, 112], which was at the other extreme against the molecular theory. Their utility in describing mechanical behaviors of elastomers is evident: For example, the use of the finite element method (FEM, mathematically a kind of the calculus of variations) needs to have numerical values representing the stress–strain behaviors of the sample. FEM is now a very powerful technique for the designing of many rubber products.

On the other hand, the molecular models proposed after the classical one have been only partially successful in terms of quantitative description of mechanical behaviors. For example, use of the inverse Langevin chain model instead of the Gaussian had to have suggested more exact description of rubber elasticity, since the Gaussian is just the simplest case of the inverse Langevin, mathematically speaking. However, use of the inverse Langevin function did not give enough betterment or improvement to counterbalance the mathematical complexity of using it. "Up to how high an extension the use of the Gaussian chain model is valid?" is a question still now to be checked by an experimental approach. On this line, one impressive paper [113] is notable, which was published in 1964. The authors of this article worked on NR vulcanizates, and checked the non-Gaussian behavior under a large deformation together with the strain-induced crystallization (SIC, see Sect. 3.4. SIC is observed above the elongation ratio of five or so) behavior. They had aimed to elucidate the reasons of the abrupt stress increase at a high strain, SIC, or fully extended rubber chains (i.e., non-Gaussian chain effect). However, the details of SIC were not known then (see Sect. 3.4), and the C2 term determined from the Mooney-Rivlin plot was not exactly a constant as later revealed (see Sect. 3.1.1). Due to the two problems, the authors' excellent efforts based on the then most updated techniques were in vain in elucidating the non-Gaussian behaviors of stretched rubber chains.

At present, several molecular theories of rubber elasticity are still under dis-cussion. One of the foci is a contribution of topological entanglements in the networks to the rubber elasticity. The Flory's school tends to disregard the entan-glements to insist on the phantom network model [86, 108, 109, 114]. From the rheological viewpoint, the contribution of entanglement is considered to be very important [110]. Urayama et al. [115–117] elucidated the importance of entanglements using cross-linked networks prepared in solution. They also used the biaxial stretching technique [118–122], which had been used in the phenomeno- logical studies of rubber mechanics, in comparing the several theories on rubber elasticity (see Sect. 3.1.1). Even this excellently revealing study is not conclusive yet in determining which rubber elasticity theory is most promising.

SIC behaviors of NR, which Ref. [113] had discussed, have recently been elucidated much by the utilization of time-resolved study at synchrotron radiation beam lines (see Sect. 3.4), and on mechanical properties, the uniaxial tensile mode has been found not enough to elucidate the mechanical behaviors at higher elon- gation (see Sect. 3.1.1). Consequently, the scientific preconditions for establishing new molecular rubber elasticity theory seem to be historically maturing now. Rubber elasticity is again to be highlighted for the further advance or innovation of polymer science as well as of rubber science.

Remark 2 Goodyear and Oenslager

It is well known that vulcanization (cross-linking reactions of rubber by using sulfur) was invented by C. Goodyear (1800–1860), an American, in 1839. However, there have been a few other stories on the invention of vulcanization. One story claimed that the inventor was T. Hancock (1786–1865), an Englishman: The first patent of vulcanization was due to Hancock. Another story puts both Goodyear and Hancock at the inventors' chair, that is, vulcanization was coinvented by recognizing Hancock's contributions to the rubber processing and vulcanization techniques.

The issue associated with vulcanization is complicated, but may be summarized as follows: Goodyear's invention was in 1839, but he had been so poor to afford the cost of patenting of his invention. He, therefore, tried to get supporters to help him to apply for the patent. In doing so, he carelessly distributed some of his samples (pieces of the rubber vulcanizates) to his supporters in order to display their superior properties, when asking for donation. Eventually, one of the samples reached Hancock's hands via his friend, a businessman, who just came back to England from USA. Hancock had been working on rubber more years than Goodyear had, and he was stunned upon touching it: He recognized the superiority of the rubber sample, and he also noticed some powdery materials on the sample surface. He perhaps instantly estimated they might be bloomed sulfur. He was absorbed in working on the sample to reproduce it in his laboratory. And, he managed to submit the patent document to the office in England, a little earlier than Goodyear. Goodyear started an action at law in England, but his money was not enough to continue the lawsuit. Thus, Hancock's English patent was finalized (June 30, 1847, No. 12,007). As a matter of formality, this is the first patent dealing with vulcan- ization [123]. Since Goodyear had enough evidences and a few witnesses for his invention in 1839, historically speaking, this invention is evidently due to Goodyear, not Hancock.

However, Goodyear's difficult years continued. He faced trials for his debts not only in USA, but also in France and in England. While Hancock was a mechanical engineer and he managed to process rubber using his own mixer and others,

Goodyear used his fingers and hands to operate knifes and a wooden hammer to work on rubber. He tried lots of chemical reagents to be mixed in rubber by using his muscular strength. Most probably, he used his bare hands and fingers when treating powdery reagents including many poisonous ones, e.g., lead compounds, which possibly was one of the causes of his death. However, he still published a book privately, which suggests that his understanding of rubber was surprisingly systematic for a street inventor [124].

Goodyear did, or was obliged to do, his work at home in kitchen. Thus, a famous anecdote, one of his favorite rubber balls (mixes of rubber with sulfur and lead white) fell down by chance to his wife's cooking stove led to his discovery of vulcanization, has somewhat the feel of truth. His ordeal, full of poverty, patience, and hard workings, were great, yet, his wife Clarissa's life, who always supported him without complaining, who managed to raised up five children by careful housekeeping, and who passed away by sickness before him, was surely more impressive and touching [125]. The popularization of the above anecdote among American people might be the result of their warm sympathy and sincere compassion on her. Their life may be a historical monument when the American capitalism was getting swing to catch up Europe, remote from but firmly on the way to Pax Americana in the twentieth century.

Single use of sulfur resulted in very low rate of vulcanization. Many metallic compounds (mainly metal oxides) showed some effects but not enough. Early in the twentieth century, an organic compound was found by G. Oenslager (1873–1956) to accelerate the vulcanization. He was graduated from Harvard as chemistry major, and at the graduate school there he was trained by Prof. T.W. Richards (winner of Nobel Prize of Chemistry in 1914). After obtaining his master's degree, he worked at a pulp and paper manufacturer.

In April 1905, Oenslager joined the Diamond Rubber Company as a research chemist. At that time, the company was working on development and manufacturing of automobile tires. He was in charge of sulfur vulcanization. He assumed that a little acceleration effects of metallic oxides were due to their weak basicity, and he looked for organic substitutes. When tried aniline as an organic base, he found much better effect on vulcanization. However, aniline is liquid (boiling point, 184 °C) and was toxic and inconvenient for mixing with rubber. He recommended thiocarbanilide as a solid base. The Ford T model was on the market in 1908, and cars were increasingly accepted by ordinary citizens. Under these social and industrial conditions, Oenslager's finding has stimulated a cutthroat competition for good organic accelerators among chemical companies, which continues up to now. He recognized the function of zinc oxide, and he laid a starting line of the vulcanization paradigm, i.e., the suitable recipes of sulfur/organic accelerator system with activators (usually zinc oxide and stearic acid) are to be designed with some other auxiliary reagents, and a few cases are regarded as more or less standard recipes which were established in the 1960s.

Oenslager was not a synthetic organic chemist, and he left the developmental work of accelerators afterward. The Diamond Company was purchased by Goodrich in 1912, and in 1920 he was dispatched to Yokohama Rubber Company in Japan for technical guidance. During his stay in Japan, he was interested in Oriental cultures and had a good impression to Japan. The origin of "the Oenslager Award" from Society of Rubber Industry, Japan, for an excellent achievement on rubber research is his stay in Japan this time. He returned to USA via Java, Sumatra, and Malay where he visited some natural rubber plantations and rubber institutions [126].

References

1. R.E. Burk, H.E. Thompson, A.J. Weith, I. Williams, *Polymerization and Its Applications in the Fields of Rubber, Synthetic Resins, and Petroleum* (Reinhold Publishing Co., New York, 1937)
2. C.Z. Duisberg, Z. Elektrochem. **24**, 369 (1918)
3. K. Gottlob, Gummi Ztg. **33**, 508 (1919)
4. C.H. Bamford, W.G. Barb, A.D. Jenkins, P.F. Onyon, *The Kinetics of Vinyl Polymerization by Radical Mechanisms* (Butterworths Scientific Publications, London, 1958)
5. Kh.S. Bagdasaryan, *Theory of Free-Radical Polymerization* (Israel Program for Scientific Translation, Jerusalem, 1968) [Original Russian edition was published in 1959]
6. J.C. Bevington, *Radical Polymerization* (Academic Press, New York, 1961)
7. G.S. Whitby, C.C. Davis, R.F. Dunbrook (eds.), *Synthetic Rubber* (John Wiley & Sons, New York, 1954)
8. J.P. Kennedy, E.G.M. Törnqvist (eds.), *Polymer Chemistry of Synthetic Elastomers* (Interscience Publishers, New York, 1968)
9. W.M. Saltman (ed.), *The Stereo Rubbers* (Wiley Interscience, New York, 1977)
10. G. Holden, N.R. Legge, R.P. Quirk, H.E. Schroeder (eds.), *Thermoplastic Elastomers*, 2nd edn. (Hanser Publishers, Munich, 1996)
11. H. Sinn, W. Kaminsky, H.-J. Vollmer, E. Woldt, Angew. Chem. Int. Ed. **19**, 390 (1980)
12. E.Y.-X. Chen, Chem. Rev. **109**, 5157 (2009)
13. S. Kaita, Z. Hou, M. Nishiura, Y. Doi, J. Kurazumi, A. Horiguchi, Y. Wakatsuki, Macromol. Rapid Commun. **24**, 179 (2003)
14. M. Kato, M. Kamigaito, M. Sawamoto, T. Higashimura, Macromolecules **28**, 1721 (1995)
15. K. Matyjaszewski, J.H. Xia, Chem. Rev. **101**, 2921 (2001)
16. M. Kamigaito, T. Ando, M. Sawamoto, Chem. Rev. **101**, 3689 (2001)
17. M. Ouchi, T. Terashima, M. Sawamoto, Chem. Rev. **109**, 4963 (2009)
18. C.W. Bielawski, R.H. Grubbs, Prog. Polym. Sci. **32**, 1 (2007)
19. L. Bateman, C.G. Moore, M. Porter, B. Saville, in *The Chemistry and Physics of Rubber-like Substances*, ed. by L. Bateman (Maclaren, London 1963), Chap. 15
20. I.R. Gelling, M. Porter, in *Natural Rubber Science and Technology*, ed. by A.D. Roberts (Oxford University Press, Oxford, 1988), Chap. 10
21. F.A. Carey, R.J. Sundberg, *Advanced Organic Chemistry, Part B: Reactions and Synthesis*, 5th edn. (Springer, Berlin, 2007)
22. P. Phinyocheep, in *Chemistry, Manufacture and Applications of Natural Rubber*, ed. by S. Kohjiya, Y. Ikeda (Woodhead/Elsevier, Cambridge, 2014), Chap. 3
23. D.R. Burfield, K.-L. Lim, K.-S. Law, S. Ng, Polymer **25**, 995 (1984)
24. A.S. Hashim, S. Kohjiya, Kautsch. Gummi Kunstst. **46**, 208 (1993)
25. P. Theato, H.-A. Klok (eds.), *Functional Polymers by Post-Polymerization Modification—Concepts: Guidelines and Applications* (Wiley-VCH, Weinheim, 2012)
26. J.E. Mark, S.-J. Pan, Macromol. Rapid Commun. **3**, 681 (1982)
27. S. Kohjiya, Y. Ikeda, Rubber Chem. Technol. **73**, 534 (2000)

28. T. Ohashi, A. Tohsan, Y. Ikeda, Polym. Int. **66**, 250 (2017)
29. A. Tohsan, Y. Ikeda, in *Chemistry, Manufacture and Applications of Natural Rubber*, ed. by S. Kohjiya, Y. Ikeda (Woodhead/Elsevier, Cambridge, 2014), Chap. 6
30. A. Kato, A. Tohsan, S. Kohjiya, T. Phakkeeree, P. Phinyocheep, Y. Ikeda, in *Progress in Rubber Nanocomposites*, ed. by S. Thomas, H.J. Maria (Woodhead/Elsevier, Amsterdam, 2017), Chap. 12
31. S. Kohjiya, K. Murakami, S. Iio, T. Tanahashi, Y. Ikeda, Rubber Chem. Technol. **74**, 16 (2001)
32. S. Poompradub, S. Kohjiya, Y. Ikeda, Chem. Lett. **43**, 672 (2005)
33. Y. Ikeda, S. Poompradub, Y. Morita, S. Kohjiya, J. Sol-Gel. Sci. Technol. **45**, 299 (2008)
34. E. Miloskovska, C. Friedrichs, D. Hristova-Bogaerds, O. Persenair, M. van Duin, M.R. Hansen, G. de With, Macromolecules **48**, 1093 (2015)
35. K. Yoshikai, T. Ohsaki, M. Furukawa, J. Appl. Polym. Sci. **85**, 2053 (2002)
36. A. Tohsan, P. Phinyocheep, S. Kittipoomc, W. Pattanasiriwisawad, Y. Ikeda, Polym. Adv. Tech. **23**, 1335 (2012)
37. A. Tohsan, R. Kishi, Y. Ikeda, Colloid Polym. Sci. **293**, 2083 (2015)
38. Y. Ikeda, A. Tohsan, Colloid Polym. Sci. **292**, 567 (2014)
39. F. Thurn, E. Meyer-Simon, R. Michel, Ger. Offen, DE 2212239 A1 19731004 (1973)
40. S. Wolff, Kautsch. Gummi Kunstst. **30**, 516 (1977)
41. S. Wolf, Rubber Chem. Technol. **69**, 325 (1996)
42. A.S. Hashim, B. Azahari, Y. Ikeda, S. Kohjiya, Rubber Chem. Technol. **71**, 289 (1998)
43. A.S. Hashim, Y. Ikeda, S. Kohjiya, Polym. Inter. **38**, 111 (1995)
44. C. Goodyear, in *Gum-Elastic and Its Varieties, with a Detailed Account of Its Application and Uses and of the Discovery of Vulcanization*. Published for the author, New Haven (1855)
45. B.K. Peirce, in *Trials of an Inventor: Life and Discoveries of Charles Goodyear* (1866) [Reprinted from the 1866 edition by University Press of the Pacific, Honolulu in 2003]
46. C.O. Weber, *The Chemistry of India Rubber, Including the Outline of a Theory on Vulcanisation* (Charles Griffin & Co., London, 1902)
47. S. Kohjiya, *Natural Rubber: From the Odyssey of the Hevea Tree to the Age of Transportation* (Smithers RAPRA, Shrewsbury, 2015)
48. G. Oenslager, Ind. Eng. Chem. **25**, 232 (1933)
49. A.Y. Coran, ChemTech **13**, 106 (1983)
50. A.V. Chapman, M. Porter, in *Natural Rubber Science and Technology*, ed. by A.D. Roberts (Oxford University Press, Oxford, 1988), Chap. 12
51. A.Y. Coran, in *Science and Technology of Rubber*, 2nd edn., ed. by J.E. Mark et al. (Academic Press, San Diego, 1994), Chap. 7
52. T.S. Kuhn, *The Structure of Scientific Revolutions* (University of Chicago Press, Chicago, 1962)
53. Y. Ikeda, H. Kobayasi, S. Kohjiya, Kagaku (Chemistry) **70**, 19 (2015) [In Japanese]
54. D.J. Elliott, in *Developments in Rubber Technology -1*, ed. by A. Whelan, K.S. Lee (Applied Science Pub., London, 1979), pp. 1–44
55. A.Y. Coran, in *The Science and Technology of Rubber*, 4th edn., ed. by B. Erman, J.E. Mark, C.M. Roland (Elsevier, Amsterdam, 2013), Chap. 7
56. Y. Ikeda, in *Chemistry, Manufacture and Applications of Natural Rubber*, ed. by S. Kohjiya, Y. Ikeda (Woodhead/Elsevier, Oxford, 2014), Chap. 4
57. Y. Ikeda, N. Higahsitani, K. Hijikata, Y. Kokubo, Y. Morita, M. Shibayama, N. Osaka, T. Suzuki, H. Endo, S. Kohjiya, Macromolecules **42**, 2741 (2009)
58. Y. Ikeda, Y. Yasuda, T. Ohashi, H. Yokoyama, S. Minoda, H. Kobayashi, T. Honma, Macromolecules **48**, 462 (2015)
59. C.S.L. Baker, in *Natural Rubber Science and Technology*, ed. by A.D. Roberts (Oxford University Press, Oxford, 1988), Chap. 11
60. P.R. Dluzneski, Rubber Chem. Technol. **74**, 451 (2001)
61. S. Kohjiya, Nippon Gomu Kyokaishi **49**, 459 (1976) [In Japanese]

62. B.G. Crowther, P.M. Lewis, C. Metherell, in *Natural Rubber Science and Technology*, ed. by A.D. Roberts (Oxford University Press, Oxford, 1988), Chap. 6

63. P. Tangboriboonrat, D. Polpanich, T. Suteewong, K. Sunguunsap, U. Paiphansiri, C. Lerthititrakal, Colloid Polym. Sci. **282**, 177 (2003)

64. Y. Matoba, Y. Ikeda, S. Kohjiya, Solid State Ionics **147**, 403 (2002)

65. S. Yamashita, A. Yamada, M. Ohata, S. Kohjiya, Makromol. Chem. **186**, 1373 (1985)

66. K. Urayama, Y. Ikeda, S. Kohjiya, Nippon Gomu Kyokaishi **68**, 814 (1995) [In Japanese]

67. S. Kohjiya, K. Urayama, Y. Ikeda, Kaut. Gummi Kunstst. **50**, 868 (1997)

68. A. Kidera, T. Higashira, Y. Ikeda, K. Urayama, S. Kohjiya, Polym. Bull. **38**, 461 (1997)

69. K. Urayama, S. Kohjiya, M. Yamamoto, Y. Ikeda, A. Kidera, J. Chem. Soc. Faraday Trans. **93**, 3689 (1997)

70. S.R. Elliott, *Physics of Amorphous Materials*, 2nd edn. (Longman, Harlow, 1990)

71. P.H. Geil, *Polymer Single Crystals* (Interscience Publishers, New York, 1963)

72. S. Onogi, *Kagakushanotameno Reoroji (Rheology for Chemists* (Kagakudojin, Kyoto, 1982) [In Japanese]

73. G.R. Strobl, *The Physics of Polymers* (Springer, Berlin, 1996)

74. M.L. Williams, R.F. Landel, J.D. Ferry, J. Am. Chem. Soc. **77**, 3701 (1955)

75. M.L. Williams, J. Phys. Chem. **59**, 95 (1955)

76. Y. Kobayashi, W. Zheng, E.F. Meyer, J.D. McGervey, A.M. Jamieson, R. Simha, Macromolecules **22**, 2302 (1989)

77. J.E. Catsiff, A.V. Tobolsky, J. Colloid Sci. **10**, 375 (1955)

78. S. Onogi, T. Masuda, K. Kitagawa, Macromolecules **3**, 109 (1970)

79. L.R.G. Treloar, *The Physics of Rubber Elasticity*, 3rd edn. (Clarendon Press, Oxford, 1975)

80. P.J. Flory, *Principles of Polymer Chemistry* (Cornell University Press, Ithaca, 1953)

81. K.H. Meyer, C. Ferri, Helv. Chim. Acta **18**, 570 (1935)

82. G.S. Rushbrooke, *Introduction to Statistical Mechanics* (Clarendon Press, Oxford, 1951)

83. R.W. Gurney, *Introduction to Statistical Mechanics* (Dover, New York, 1966)

84. D.H. Trevena, *Statistical Mechanics: An Introduction* (Woodhead Publishing, Oxford, 2010)

85. F. Bueche, *Physical Properties of Polymers* (Interscience Publishers, New York, 1962)

86. B. Erman, J.E. Mark, *Structures and Properties of Rubberlike Networks* (Oxford University Press, Oxford, 1997)

87. C.M. Roland, *Viscoelastic Behavior of Rubbery Materials* (Oxford University Press, Oxford, 2011)

88. Y. Furukawa, *Inventing Polymer Science: Staudinger, Carothers, and the Emergence of Macromolecular Chemistry* (University of Pennsylvania Press, Philadelphia, 1998)

89. H. Staudinger, *Die Hochmolekularen Organischen Verbindungen—Kautschuk und Cellulose—* (Springer, Berlin, 1960) [Original edition was published in 1932]

90. H. Staudinger, *Arbeitserrinerungen* (Dr. Alfred Hüthig Verlag, Heidelberg, 1961)

91. H. Mark, G.S. Whitby (eds.), *Collected Papers of Wallace Hume Carothers on the High Polymeric Substances* (Interscience, New York, 1940)

92. J.R. Katz, Naturwiss **13**, 410 (1925)

93. Houwink, in *Elasticity, Plasticity and Structure of Matter* (Cambridge University Press, Cambridge, 1937), Chap. 7

94. K.H. Meyer, H. Mark, Ber. Dtsch. Chem. Ges. **61**, 1939 (1928)

95. G. Patterson, *A Prehistory of Polymer Science* (Springer, Heidelberg, 2012)

96. K.H. Meyer, C. Susich, E. Valko, Kolloid Z. **59**, 208 (1932)

97. W. Kuhn, Kolloid Z. **68**, 2 (1934)

98. W. Kuhn, Kolloid Z. **76**, 258 (1936)

99. W. Kuhn, Kolloid Z. **87**, 3 (1939)

100. E. Guth, H.F. Mark, Monatsh. Chem. **65**, 93 (1934)

101. F.T. Wall, J. Chem. Phys. **11**, 67 (1941)

102. E. Guth, H.M. James, Ind. Eng. Chem. **33**, 624 (1941)

103. L.R.G. Treloar, Trans. Faraday Soc. **39**, 241 (1943)

104. H.M. James, E. Guth, J. Chem. Phys. **11**, 455 (1943)

105. P.J. Flory, Chem. Rev. **35**, 51 (1944)
106. R. Kubo, *Gomu Dansei (Rubber Elasticity)* (Kawade Shobo, Tokyo, 1947) [In Japanese]
107. F.W. Billmeyer, Jr., *Textbook of Polymer Chemistry* (Interscience Publishers, New York, 1957) [Later, the title was changed to *Textbook of Polymer Science*]
108. B. Erman, J.E. Mark, C.M. Roland, *The Science and Technology of Rubber*, 4th edn. (Elsevier, Amsterdam, 2013)
109. J. Mark, K. Ngai, W. Graessley, L. Mandelkern, E. Samulski, J. Koenig, G. Wingnall, *Physical Properties of Polymers*, 3rd edn. (Cambridge University Press, Cambridge, 2004)
110. W.W. Graessley, *Polymer Liquids and Networks: Dynamics and Rheology* (Garland Science, London, 2008)
111. S. Kawabata, H. Kawai, Adv. Polym. Sci. **24**, 89 (1977)
112. G. Saccomadi, R.W. Ogden (eds.), *Mechanics and Thermomechanics of Rubberlike Solids* (Springer, Wien, 2004)
113. K.J. Smith, Jr., A. Greene, A. Ciferri. Kolloid Z. Z. Polym. **194**, 49 (1964) [This journal is the second generation of *Kolloid Z.* as seen in Refs. 96–99. Now, title of the third generation is in English, *Colloid and Polymer Science*, accepting manuscripts both in German and in English]
114. B. Erman, in *Mechanics and Thermomechanics of Rubberlike Solids*, ed. by G. Saccomadi, R.W. Ogden (Springer, Wien, 2004), pp. 63–89
115. K. Urayama, S. Kohjiya, J. Chem. Phys. **104**, 3352 (1996)
116. K. Urayama, T. Kawamura, S. Kohjiya, J. Chem. Phys. **105**, 4833 (1996)
117. T. Kawamura, K. Urayama, S. Kohjiya, Nihon Reoroji Gakkaishi **25**, 195 (1997)
118. T. Kawamura, K. Urayama, S. Kohjiya, Macromolecules **34**, 8252 (2001)
119. K. Urayama, T. Kawamura, S. Kohjiya, Macromolecules **34**, 8261 (2001)
120. T. Kawamura, K. Urayama, S. Kohjiya, J. Polym. Sci., Part B: Polym. Phys. **40**, 2780 (2002)
121. K. Urayama, T. Kawamura, S. Kohjiya, J. Chem. Phys. **118**, 5658 (2003)
122. T. Kawamura, K. Urayama, S. Kohjiya, Nihon Reoroji Gakkaishi **31**, 213 (2003)
123. T. Hancock, *Personal Narrative of the Origin and Progress of the Caoutchouc or India-Rubber Manufacture in England* (Longmans & Roberts, London, 1857)
124. C. Goodyear, *Gum-Elastic and Its Varieties, with a Detailed Account of Its Application and Uses and of the Discovery of Vulcanization* (New Haven, 1855) [Published for the author]
125. C. Slack, *Noble Obsession: Charles Goodyear, Thomas Hancock, and the Race to Unlock the Greatest Industrial Secret of the Nineteenth Century* (Hyperion, New York, 2002)
126. B.N. Zimmerman (ed.), *Vignettes from the International Rubber Science Hall of Fame 1958–1988: 36 Major Contributors to Rubber Science* (Rubber Division of American Chemical Society, Akron, 1989), pp. 170–177

Chapter 3
Materials Science of Rubber

3.1 Beginning and Development of Materials Science

As already described in Sect. 1.1, materials are as old as human beings, since the first humans were assumed to evolve from the ape by using simple tools prepared from stones and/or woods for hunting and gathering, followed by the utilization of the bronze and the iron. Through the classical (Greek and Roman) civilizations and the medieval periods, the use of iron had been spread more or less worldwide [1–3]. Mining and metallurgical industry abounded in the 2nd half of the medieval period, and some famous and scientifically excellent books were published [3, 4]: For example, Georgius Agricola (1494–1555) authored a book entitled *De Re Metallica* [5], which describes techniques of ore mining and extraction of metal from the ore in details with lots of figures by wood prints. The description was very systematic, yet it remained to be an extensive explanative description, having failed to give rise to the way to establishing metallurgy as a modern technology. For instituting technology as a branch of science, two historical preconditions were to be fulfilled: One, the modern scientific features, especially use of mathematics and well-planned experimental methods, were put on a firm footing through the Scientific Revolution [6–10], during which physics was first established as the modern science, followed by chemistry [7–12]. Secondly, since technology is a science on technics (see Table 1.1), the spectacular progress of technics through the Industrial Revolution was the requirement for the birth and progress of technology as a branch of science [9, 13, 14]. One more important issue was already described in Sect. 1.1.3: During the period from the middle to the end of the nineteenth century, higher educational colleges or universities on technology had been established in Germany, USA, and a little later in Japan, and technology as a branch of science of technics was extensively taught there to result in the social recognition of technology early in the twentieth century.

In Ref. [14], Chapter 6 is entitled "Materials of Modernism, 1890–1950," which suggests that materials science was established at about 1900 or so. Also, Ref. [15]

© Springer Nature Singapore Pte Ltd. 2018
Y. Ikeda et al., *Rubber Science*, https://doi.org/10.1007/978-981-10-2938-7_3

briefly describes "The Beginning of Materials Science," in which the origin of materials science was assumed to be the study on crystalline ferrite (from the Latin for iron) and pearlite (a mixture of pure iron crystals and carbon-rich crystals) by H. C. Sorby, a British geologist and microscopist. Near the end of the nineteenth century, he observed the crystalline structure of steel by polishing the surface and etching it with acid to discover ferrite and pearlite, and opened the door to elucidating the relationship among processing, structure, and property of steels, ultimately having led to the much improvement and mass production of steels [15]. Further, two Nobel prize-winning studies in physics, i.e., discovery of X-ray diffraction by crystalline materials by M. von Laue in 1914 and structural elucidation by X-ray diffraction by W.H. and W.L. Braggs in 1915, gave rise to a quantum leap of materials science: In the 1920s, use of X-ray much accelerated the structural studies on materials (see Sect. 3.4.1), which continues to be the case up to now. After the 1950s, the transmission electron microscope (TEM) became popular in research laboratories including those at research and development facilities instituted at companies, followed by the scanning electron microscope (SEM), the Auger spectrometer, the scanning tunneling microscope (STM), and so on. They have been installed at many materials research laboratories for micro- and nanostructure determination of various materials, to result in the maturing of materials science in revealing the structure–property relationships. Without the elucidation of structure-property relationships, many important devices, for example, diode and transistor would not have been invented, hence many atoms, e.g., silicon, germanium, arsenic, would not have found their place in the material science textbooks. Also, modern preparative techniques, such as molecular beam epitaxy, chemical vapor deposition, sol-gel chemistry, have enabled building materials atom by atom. The most recent assignment on materials science may be to prepare the new materials the properties of which are theoretically predicted based on the material and device designs. In other words, new materials are being produced not by the repetitive trial-and-error method, but in accordance with the rationally prescribed design. This is likely to be one of the factors giving rise to the next stage of materials science in a near future.

It is still under active discussion that the developmental process of industrial technics and technology is revolutionary (discontinuity is emphasized) [6–11, 16–18] or evolutionary (continuity is predominant) [2, 11, 13, 14, 18–20] in the history and philosophy of science and technology arena. Among these, literatures 11 and 18 are featuring both pros and cons on the two, from a historiographical standpoint. The result of this debate may influence on the historical description given above. It is notable here again that, in majority of these books, technology, and technics are used interchangeably or technology includes both technics and a scientific discipline on technics, while technology means not technics but a science of technics (see Table 1.1, again) here. In this meaning, a few authors have preferred engineering to technology, since the full-scale development of engineering has started during the Industrial Revolution and it is still the driving power of industrialization.

3.2 Physical Properties of Materials

Various properties of materials are so far broken down into two: physical properties and chemical properties. Recently, however, biomaterials are increasing its importance, and biological properties are equally important now. However, all materials are requested to have a certain mechanical stability under the serving conditions, together with some inertness in chemical and biological meanings, in general. The so-called highly functional materials, which become more and more important for the modern society, are chemically and biochemically active, and their function needs to work more or less under dynamic conditions for their functioning. Development of those highly functionalized or bioactive materials is now a driving force for the renewed advance of materials science. Materials science of the twentieth century might have been matured as one of the modern sciences at the third quarter of the twentieth century as described briefly in Sect. 3.1. Additionally, it has to be emphasized the importance of structure determination of materials including the devices there from. Actually, without knowing the structure, the progress in materials had been sluggish, since the modern Materials Age has been due to the establishment of materials science, structure–property relationship in particular, in the twentieth century.

In this chapter, therefore, brief summary of the development of materials science in general is given at first as above, and some examples of preparation and characterization techniques necessary to develop new soft materials are described from a viewpoint of materials science. For the general and practical properties of rubbery materials on the market, refer to the table in Appendix.

3.2.1 Mechanical Properties

Usefulness of materials is firstly based on their physical property, and hence, the essence of materials science has been concerned with the physical properties of materials. Among many physical properties, the most valued has been mechanical properties, since it is the a priori precondition that all the materials must be mechanically stable under the serving conditions. When a new material is developed, it must keep its dimensions, however excellent its other properties might be. Otherwise, the so-called carrier or supporter has to be used for carrying the material. In this case, the mechanical properties of the supporter are to be checked, of course.

(1) Uniaxial Tensile Measurements

For rubbery materials, the first step of mechanical property evaluation is a uniaxial tensile measurement to show the tensile stress–strain (S-S) behavior, often by stretching up to the ten time length of the original one. Popularly, this extraordinary high extensibility is the most impressive character of rubber. Industrially, the predominant importance of uniaxial tensile S-S measurements is dependent on the quality control of rubber products to be on the market in accordance with the ISO standards. Among so many mechanical properties of rubber, the uniaxial tensile measurement is very impressive to understand the surprisingly high extensibility of rubber, but it is just a first step from scientific point of view. The measurements by ISO (or ASTM, BA, DIN, JIS, and any other equivalents) provisions are for quality control of the industrial products and not much of use for scientific elucidation of the physical nature of rubbery substances: The results of ISO tests are not an observation under a quasi-static process of thermodynamic meanings. Once upon a time, a famous Japanese physicist wrote [21], "It is quite regrettable that a mountain of data of mechanical (tensile) test results having been accumulated on rubber products are useless due to the insufficiency of experimental conditions for the scientific studies, in particular." This suggestion has not been accepted much by rubber chemists and engineers except one paper [22] as far as the authors know: K. Ahagon reported that only 100% modulus (stress at 100% elongation) was in harmony with the classical rubber elasticity theory (see Sect. 2.4.2) among the huge number of tensile results of rubber goods measured per the JIS or ASTM provisions. He has, thereafter, developed his tensile behavior discussions on rubber mainly using 100% modulus in order to make full use of the reported data for the scientific discussions [22–25].

From a scientific viewpoint, it is emphasized that the important first step for the mechanical characterization of rubbery materials is uniaxial tensile measurements under as near conditions as possible to the quasi-static process. Practically speaking, to check the effect of deformation rate of each rubber sample at room temperature has to be conducted to determine a suitable stretching speed at first. The rate specified in the ISO (ASTM, JIS and the other) standards are definitely too high. As a compromised yet possible example, 10 mm per minute may be a recommendable deformation rate in the scientific studies. If you stick to uniaxial tensile tests further in order to make fully scientific discussions, you are to have the results obtained at several deformation rates at various temperatures. Or, you had better consider the multiaxial tensile and/or dynamic mechanical measurements described below.

(2) **Multiaxial Tensile Measurements: Phenomenological Approach**

Figure 3.1 shows the physically possible coverage of the general biaxial mode of tensile measurements, under an assumption of incompressibility of rubber, i.e., $I_3 = (\lambda_1 \lambda_2 \lambda_3)^2 = 1$ [26]. Here, "incompressibility" means no volume change upon deformation. The coverage of uniaxial, equi-biaxial, and pure shear is also shown in the figure. On the X-axis and Y-axis, see Eqs. (3.2), (3.3), (3.6), and (3.10) to be shown later.

As shown in this figure, the usual uniaxial deformation covers mechanical information only at the edge of the possible area. Therefore, the experimental results at the edge are not sensitive enough to consider the mechanical properties of rubber, e.g., to tell apart among the possible rubber elasticity theories. A typical example of such a case is the overestimation of the Mooney–Rivlin plot of uniaxial data [27],

Fig. 3.1 Possible covering area by the general biaxial mode of mechanical deformation compared with that of uniaxial, equi-biaxial, or pure shear (from Fig. 1(b) in Ref. [26])

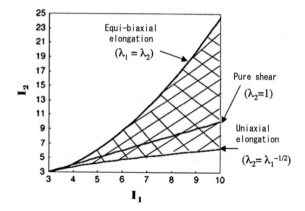

which is explained later. Among the biaxial deformations, the equi-biaxial covers also only information at the upper side edge. Only general biaxial (λ_1 and λ_2 are independently changed) covers all the possible area; hence, the general biaxial measurements are highly recommendable for the tensile behavior elucidation of any elastomers. However, the biaxial measurement necessitates a more expensive tensile tester and takes much more time than the uniaxial one. Consequently, some researchers recommend pure shear as a substitute of the general biaxial mode [28]: The pure shear (biaxial but one axis is kept at a constant length) is covering not the edges but just the middle of the inner part as shown in Fig. 3.1, which may somewhat justify their claim. From the experimental point of view, the fixing of the rubber sample at the movable holders is still to be improved in accordance with the very high elongation of rubber.

Biaxial deformation behaviors are analyzed by the so-called phenomenological theory of rubber elasticity [26–31]. By deforming a rubber sample which is a homogeneous, isotropic and elastic material, elastic energy is accumulated in the sample, which is expressed by strain energy density function, W.

$$W = W(I_1, I_2, I_3). \tag{3.1}$$

That is, W is the function of the invariants of the Green's deformation tensor I_i ($i = 1, 2, 3$). The invariants are not varied by the axis of coordinates. The definitions of them are

$$I_1 = \lambda_1^2 + \lambda_2^2 + \lambda_3^2, \tag{3.2}$$

$$I_2 = \lambda_1^2\lambda_2^2 + \lambda_2^2\lambda_3^2 + \lambda_3^2\lambda_1^2, \tag{3.3}$$

$$I_3 = \lambda_1^2\lambda_2^2\lambda_3^2, \tag{3.4}$$

where λi (i = 1, 2, 3) are the deformation ratio at each principal axis. The third invariant is equal to unity for incompressible materials, that is,

$$I_3 = (\lambda_1 \lambda_2 \lambda_3)^2 = 1. \tag{3.5}$$

And Eq. (3.1) may be reduced to

$$W = W(I_1, I_2). \tag{3.6}$$

The W is given as follows by the phenomenological considerations

$$W(I_1, I_2) = \sum_{i,j=1}^{\infty} C_{ij}(I_1 - 3)^i (I_2 - 3)^j, \tag{3.7}$$

where say C_{00} = 0 in order to make the strain zero ($I_1 = I_2 = 3$) before the deformation. Using this equation allows us to evaluate the mechanical behavior of the sample once we know all the indexes C_{ij} in Eq. (3.7).

We can calculate compression, shear as well as tensile behavior of the sample by using Eq. (3.7), once we have determined all the indexes from the biaxial experiments. By the mechanical measurement, it is usual to detect force (i.e., stress), not energy, and stress is obtained by the differentiating energy by strain. Therefore, we need to have $\partial W/\partial I_i$ (i = 1, 2) for comparing with the experiments, and the following equations are to be used,

$$\partial W/\partial I_1 = \beta \left\{ \lambda_1^3 \sigma_1 / \left[\lambda_1^2 - (\lambda_1 \lambda_2)^{-2} \right] - \lambda_2^3 \sigma_2 / \left[\lambda_2^2 - (\lambda_1 \lambda_2)^{-2} \right] \right\}, \tag{3.8}$$

$$\partial W/\partial I_2 = -\beta \left\{ \lambda_1 \sigma_1 / \left[\lambda_1^2 - (\lambda_1 \lambda_2)^{-2} \right] - \lambda_2 \sigma_2 / \left[\lambda_2^2 - (\lambda_1 \lambda_2)^{-2} \right] \right\}, \tag{3.9}$$

where $\beta = 1/2(\lambda_1^2 - \lambda_2^2)$. For the sake of comparison of the equations with experimental values,

$$W = W(\lambda_x, \lambda_y) \tag{3.10}$$

is used here. When the function W is determined, stress may be calculated as a differential of strain, that is, we now know the stress–strain relations in general. One of the results obtained on end-linked poly(dimethyl siloxane) (PDMS), a model polymer network, is shown in Fig. 3.2 [31]. σ_x and σ_y are the stresses along X-axis and Y-axis, respectively. The broken line in the figure is a trace for the readers' eye guide. Triangles in the figure show the results of uniaxial, squares are those of equi-biaxial, both of which are at the lower and higher edges of possible deformation area. The perpendicular solid line is of shear deformation, which is covering the central region. It is well recognized that the general biaxial measurements are needed to cover all the deformable regions.

Fig. 3.2 Stress–strain
relationship by general biaxial
measurement of end-linked
poly(dimethyl siloxane) (from
Fig. 2 in Ref. [31])

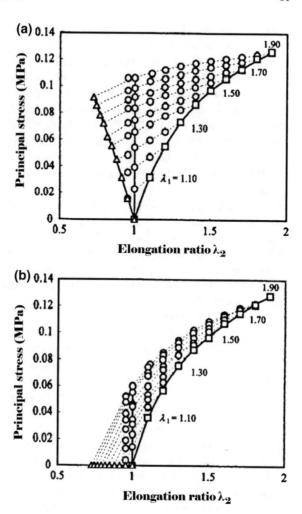

The matter in hand is to conduct such a general biaxial measurement to determine the energy (stored energy in an elastic body upon deformation) density function W of each sample. Currently speaking, the finite element method (FEM) is extensively employed in rubber industry in order to analyze the state of stress in the sample body. The FEM is a mathematical method remarkably suitable for computerization, but still its accuracy is dependent on the function W. Since the general form of W of rubbery products is not proposed yet, and we have to determine it inductively, i.e., from the biaxial experimental results. If enough number of such examples is accumulated to result in the general function form of W on rubber in a near future, we may calculate stress deductively. Holding this ultimate goal in mind, the resultant energy density function W of the end-linked PDMS model networks has been discussed in comparison with several theoretical models of rubber elasticity [32–34]. However, a definite conclusion on which theory is preferable to reproduce the biaxial tensile results has not been obtained still now. More accumulation of the biaxial data on the model networks is to be reported in a near future.

It is a repetition, but the uniaxial tensile results (plotted in triangle in Fig. 3.2) are not enough to discuss and theoretically check the mechanical behavior: The discussion based only on these uniaxial results is something like "debating the whole forest by checking one tree" [27]. The Mooney–Rivlin plot [35, 36] has afforded two parameters, C_1 and C_2, from uniaxial tensile measurement. Determination of network-chain density from C_1 may be not unreasonable, but the physical meaning of C_2 has long been discussed actively even though the constancy of it is only apparent [26–29, 37, 38]. The laborious debates, particularly those at molecular-level interpretation of C_2, have been found to be of no use. So far they have failed in finding any theoretical molecular basis, one example of which is mentioned earlier in Sect. 2.4 [39]. We have to discontinue such a discussion now, in which molecular considerations were extracted from the value (C_2) obtained by uniaxial measurements by using the phenomenological equation (the Mooney–Rivlin plot). Any molecular approaches have to be done based on the biaxial results [31–34], and further investigations on this line are to elucidate a few characteristics of modern theory of rubber elasticity [40].

While biaxial tensile elongation should be unfavorable to strain-induced crystallization (SIC, see Sect. 3.4.2; orientation to the stretching direction is the initial step of SIC), there reported a result of SIC under an equi-biaxial tensile deformation [41]. It is reasonable that the rubber chains under biaxial deformation are difficult to crystallize, since the orientation to one specific direction might be of low probability. If the reported result is not an artifact, SIC has to be taken into account during the biaxial measurements of natural rubber (NR). Check of the description in the original paper [41] suggests that the result is due to an experimental defect. That is, the author employed a cross-shaped NR sample, and the fixing of the sample to each axis was only at one point (i.e., totally fixing only at 4 points). After the biaxial deformation of the crossed sample, some regions of the sample around the fixing point seemed to be subjected to uniaxial stress to undergo SIC. From the photograph, the local deformation there is most probably estimated to be over 400% (though the overall deformation was much smaller), which is enough for SIC. SIC was not recognized at all around the central cross-part, where biaxial strain was appropriately observed. These findings strongly suggest that the shape of the subjected sample and how to fix the sample to the machine are crucial for biaxial mechanical measurements.

Recently, a polymeric gel sample (cross-linked polymer swollen by solvent) was subjected to general biaxial deformation to measure the stress relaxation behavior [42]. Not only rubbers but also soft materials in general are excellent targets of the multiaxial mechanical experiments, which is expected to expand the possibility of soft matters in a future.

(3) **Dynamic Mechanical Measurements**

Dynamic mechanical analysis (DMA) is one of the informative analytical tools for materials and is experimentally to record stress and strain as a function of temperature or time (frequency). DMA may afford much more information of use than uniaxial tensile measurements: Measuring uniaxial tensile behavior at various temperatures and at various frequencies would take too much time to give a definite conclusion. One more advisable feature of DMA compared with uniaxial tensile measurements is that DMA may give the effect of viscosity much more directory than the tensile. Rheological viewpoint indicates even elastic rubber is a viscoelastic body as a polymeric material. Hysteresis loss measurements by a cyclic uniaxial tensile mode were once popular, but the measurements at a wide temperature and frequency range are awfully time-consuming, and the interpretation is not much straightforward. Considering that rubber as a viscoelastic body has various end uses such as shock absorbers, vibration controllers, seismic isolators, and soundproofs, elucidation of dynamic viscoelastic properties of rubber is one of the most important issues for its applications. For this purpose, DMA is most convenient and informative to be given the first preference. Additionally, DMA is already an established technique: Ferry et al. published a series of papers on the cross-linked rubber in 1963 [43], and the fruit of his many years' studies was published to be regarded as a standard textbook of viscoelasticity [44].

In practice, measurements of temperature dispersion are much convenient: By using a temperature-variable instrument, a range from -160 °C (using liquid nitrogen for low-temperature regions) up to around 200 °C may be covered. Obtainable parameters are, complex dynamic modulus E^* (tensile mode) or G^* (shear mode), the real part of which are E' or G' and unreal part E'' or G'', respectively, and loss tangent $\tan \delta$, where real parts represent the elastic contribution and unreal parts and loss tangent represent the viscosity contribution to the dynamic mechanical properties of rubber. For example, the definitions of E^*, E', E'', and loss tangent are shown by the following equations.

$$E^* = E' + iE'', \tag{3.11}$$

$$E' = E^* \sin \delta, \tag{3.12}$$

$$E'' = E^* \cos \delta, \tag{3.13}$$

$$\tan \delta = E''/E'. \tag{3.14}$$

Similar relationship is defined for G^*, too. From these equations, only two are independent among E', E'', and $\tan \delta$. In other words, when we know any two values, the rest is calculated by using one of the three Eqs. (3.12–3.14). Equation (3.14) suggests loss tangent is a ratio of viscosity to elasticity, that is, a relative contribution of viscosity is evaluated as a function of temperature or frequency, which is of great use for designing vibration-related rubbery devices, including automobile tires. In the previous section, the end-linked PDMS was described.

Using a similar preparation method, cross-linked PDMS containing heterogeneous network regions was purposely synthesized, and subjected to DMA measurements. The results show relatively high tangent loss value of 0.3 at a wide temperature range between −30 °C and +150 °C [45]. This result is of use in designing rubber device for controlling vibration or soundproofing.

Among so many organic materials, low heat resistance (which is associated with low thermal conductivity, too) is the most unfavorable character of rubber in general. The origin of tire's, especially tread rubber's, essential functions for safe yet smooth driving of the car, i.e., grip, traction, wear, are originated from the friction of tread rubber with road surface. The friction is necessarily followed by heat generation resulting in the temperature increase during driving. Therefore, the evaluation of loss tangent which is related to heat generation has been the most important issue for designing tire rubber composites. For instance, skid is a movement of an automobile even after the standstill of its tires (by the driver's application of the brakes), and antiskid function of the tires especially when driving in the rain is of utmost importance (see Chap. 5). For the safe driving, the moving distance by wet-skid (a kind of sliding of tires on wet road surface) should be minimized, of course. For this issue, design of the tread rubber (which is in contact with the road surface) is essential.

Results shown in Fig. 3.3 suggest that the wet-skid resistance (relative values) is correlated with temperature at the maximum value of tangent loss of tread rubbers, which was measured at 10 Hz [46]. The samples are the 18 kinds of rubber compounds. The 4 kinds of rubber used are e-SBR, s-SBR, EPDM, and CIIR. There appear two issues to be considered on the results: First, peak temperature of tangent loss is generally in accordance with glass transition temperature T_g (see Sect. 1.3.2 and 2.3.1), but tan δ here is the case or not, and secondly, the frequency 10 Hz is appropriate or not for skidding tires, are still to be discussed. The tire tread is a typical composite, and its T_g may much dependent on how T_g is determined together with the structural details of the tire. On the latter, Roland claimed the

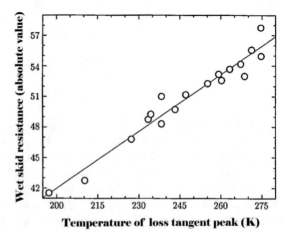

Fig. 3.3 Relationship between wet-skid resistance and the temperature at which the maximum value of tangent loss of tread rubbers was observed. Tangent loss was measured at 10 Hz. (From Fig. 7 in Ref. [46])

higher frequency may be better for simulating the skidding [47]. Anyway, wet-skid resistance, which is much complicated but very important performance for the safety issue, is sure to have somehow related to dynamic loss, and further study is highly expected. DMA as a branch of rheology is a subject to be more and more carefully studied by rubber scientists and rubber engineers.

3.2.2 Thermal Properties

(1) Importance of Thermal Properties of Rubber

The International Confederation for Thermal Analysis and Calorimetry (ICTAC) defines the heat analysis as "a group of techniques in which a property of the sample is monitored against time or temperature while the temperature of the sample, in a specific atmosphere, is programmed" [48, 49]. In accordance with this definition, the main thermal analysis methods are listed in Table 3.1 [49].

On the focused physical properties column, the first and second lines are focused on mass and volume as the basic quantities of the substance, respectively. At the first line, thermogravimetric analysis (TG) is widely used for detecting the weight change of a sample. When the rubbery sample contains inorganic filler which is quite common, the weight % of inorganic filler in it is easily quantified by TGA. Moreover, TG combined with the other numerous analysis methods has been developed. For example, in TG-DSC, TG is combined with differential scanning calorimetry (DSC), TG-MS is combined with mass spectrometry (MS), and TG-FT-IR is combined with Fourier-transform infrared spectroscopy (FT-IR). They are now general-purpose analytical instruments and are an excellent analytical method of both vulcanized rubber products and rubber compounds (a mixture of rubber before vulcanization mixed with various compounding agents, such as filler, curing agents, processing aids). The second line concerning volumetric change for rubbers fails to show any examples on rubber, one reason of which may be related to the incompressibility of rubber. However, to consider applying this technique to rubber products could possibly be an important characterization method in a future, since thermal expansion of rubber is surely different from lots of ingredients. One reference [50] reported an example on a thin polymer film.

The third and fourth lines include many examples of electroconductive, piezoelectric, magnetic, or piezomagnetic rubbers: They are usually filled with specified functional inorganic fillers. In most cases, the obtained results are very valuable since not obtainable by other measurement techniques as well as these functional rubbers are uniquely applicable materials. The fifth line focuses on the measurements of mechanical properties and the temperature variance measurements of complex modulus of elasticity, which were explained in the preceding section, are in this classification. At the last line, the calorimetric measurement (i.e., TG-DSC, TG-DTA) which was combined with TG is an indispensable method to investigate temperature change in much more detail than use of only TG.

Table 3.1 Main thermal analyses methods

Focused physical properties	Measured internal energy density (w, q)	Analysis method	Related special method
Mass (weight)	...	Thermogravimetric analysis (TGA or TG-)	TG-MS, TG-FT-IR, TG-DSC
Volume expansion coefficient or linear expansion coefficient	...	Thermodilatometry	
Dielectric constant or electric conductivity (generally, relationship between electric flux and field)	$dw = EdD$ or $dw = \Phi dp$ E = Electric field, D = Electric flux, Φ = Electric potential, p = Charge density	Dielectric thermal analysis (DETA), Thermal conductivity measurement	Thermally stimulated depolarization current measurement
Magnetic permeability (generally, relationship magnetic flux and field)	$dw = HdB$ H = Magnetic field, B = Magnetic flux	Magnetic thermoanalysis (MTA) or Thermomagnetometry	
Elastic modulus (generally, relationship between mechanical stress and strain)	$dw = (1/2)Tr[\sigma(\lambda^T)\text{-}1d(\lambda^T\lambda)\lambda^{-1}]$ σ = Stress tensor, Tr = Diagonal partial sum, λ = Deformation gradient tensor, T: Transposition	Thermomechanical analysis (TMA), dynamic mechanical analysis (DMA)	
Specific heat	$dq = Tds$	Differential thermal analysis (DTA)	Temperature
Latent heat	T = Absolute temperature s = Entropy density	Differential scanning calorimetry (DSC)	Modulation-type DSC

From Table 9.1 in Ref. [49]
MS Mass spectrometry: *FT-IR* Fourier-transform infrared spectroscopy

It is expected that the technique of the thermal measurements combined with other analytical methods is still developing. There have published a good textbook [48] about these thermal analyses, and a short review is disclosed [49]. A few recent results of thermal properties in rubber science are introduced next.

(2) Coefficient of Linear Thermal Expansion of Carbon Black-Filled Natural Rubber

The carbon black (CB)-loaded rubber vulcanizate is in fact one of the most used industrial soft materials. Among them, CB-filled NR in particular is widely used in tires for aircrafts and heavy-duty automobiles, conveyor belts, vibration isolation and seismic isolation rubber bearings, and rubbery shock absorbers. Tires are indispensable for transportation industry, and the rubber belts and the others are widely used in lots of industrial devices of manufacturing and construction as well as various commercial electronics and household appliances. The results obtained from studies on linear thermal expansion of CB-filled NR [51–54] are reviewed.

Figure 3.4a shows absolute temperature (T) dependence of thermal expansion (L) of CB-filled NR vulcanizates, whose CB loading is changed from 0 to 80 phr (designated as CB-0 to CB-80). L is defined by L (%) = 100 ($\Delta l/l_0$) where l_0 is the initial length of the sample and Δl is the change of the length by heating. L increased with T almost linearly up to 348 K. At the higher temperature region than 348 K (75 °C), the dependence of L on T became nonlinear, that is, the slope decreased little by little. And, after a maximum value L tended to remain more or less constant. In the cases of CB-0 and CB-10, L even decreased, which may suggest the beginning of thermal decomposition of NR after the maximum of L (at about 130 °C) even under a gentle nitrogen flow.

Comparison of the CB loading amount shows that the larger is the loading, the smaller is the temperature dependency, which is reasonable since the thermal change of rubber is much larger than CB. The slope of linear part between 298 and 348 K in Fig. 3.4a, namely the coefficient of thermal expansion (CTE), was cal-culated and is plotted against CB amount in Fig. 3.4b. In the figure, two CB regions are recognized, that is, High CTE region and low one, and CB-30 is near at the border. This result suggests that in the region of CB loading higher than 30 phr, the aggregation state of CB is different from that of lower CB loading samples. In the next sub-subsection (2) and in the next chapter at Sects. 4.1 and 4.2, CB networking in the rubbery matrix is introduced. The network structure forming in the rubbery matrix at the higher CB loading samples is the reason of observing the low CTE

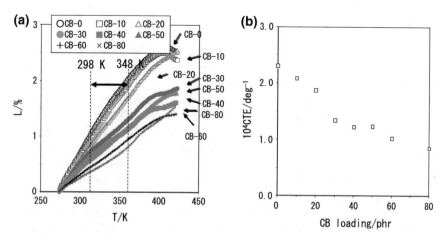

Fig. 3.4 Thermal expansion of CB-filled NR vulcanizates in the temperature region of 298–348 K under a gentle nitrogen flow. **a** Thermal expansion (L) (from Fig. 4 in Ref. [52]). **b** Change of coefficient of thermal expansion (CTE) (from Fig. 17.3 in Ref. [53])

region in Fig. 3.4b. At around 30 phr, CB aggregation develops into agglomeration (further aggregation) to form CB network structure. The CB network is assumed to restrain the increase of CTE [52, 55].

(3) Thermal Stability of Carbon Black-Loaded Natural Rubber Vulcanizate

As will be described in Sects. 4.1 and 4.2, CB in rubber, specifically here in NR, is aggregated. Additionally, at the higher loading region, CB is in the state of network structure as just positively proved later. Therefore, thermal rearrangement and collapse of the network structure of CB, which is consisting of CB aggregates in rubber matrix, are investigated by DSC and thermomechanical analysis (TMA), and these results are interpreted by the change of CB networks elucidated by three-dimensional transmission electron microscopy (3D-TEM) described at Sect. 4.1.3.

In this study [52, 55], the appropriate conditions (of temperature and atmosphere) are set up so that the rubbery matrix would not be thermally deteriorated and decomposed, i.e., practically no chemical reactions of them. The heat flow values were obtained from DSC measurements of the NR samples of 0–80 phr CB (CB-0 to CB-80) carried out under nitrogen atmosphere and at the heating rate of 10 °C/min. Figure 3.5a indicates the dependence of the flow rate on temperature. The flow rate was increased with increase of CB loading. The glass transition temperature at about 200 K and the thermal decomposition temperature at about 480 K were recognized. Unless the thermal degradation and decomposition occurred, to measure to the possible highest temperature was desirable. Experimentally, heat treatment of a number of samples at a high temperature for many hours is not convenient for an inert gas flow control; the heat treatment conditions under vacuum at 333 K (110 °C) was chosen: CB-50 and CB-80 were placed under these conditions. Figure 3.5b shows the dependence of the volume resistivity (ρ_v) of CB-50 and CB-80 on the heat treatment time (T_h) under vacuum at 333 K. The ρ_v of CB-50 was increased rapidly with the increase of the heat treatment time up to 24 h and was decreased or remained almost constant after 24 h. On the other hand, ρ_v of CB-80 was independent of the heat treatment time and maintained an almost constant value. These results revealed that the thermal stability of the CB network was in order; CB-80 $\gg$ CB-50 [52, 55].

Table 3.2 shows the results obtained from the analyses of the CB networks in CB-50 and CB-80 before and after heat treatment. For the 3D-TEM images and the visualized CB network structures of CB-50 and CB-80, refer to Refs. [52, 55]. It is suggested that even after the heat treatment complete scission of CB networks in the two samples has not observed; hence, the isolated free chains were not formed. Consequently, the coarseness and partial defects in the CB networks might have occurred by the partial scission during the heat treatment. In order to compare the CB network structures before and after heat treatment, the average lengths (L_{cross} and L_{branch}) of cross-linked and branched chains, the density (D_{cross}) of the cross-linked points, and the fractions (F_{cross} and F_{branch}) of cross-linked and

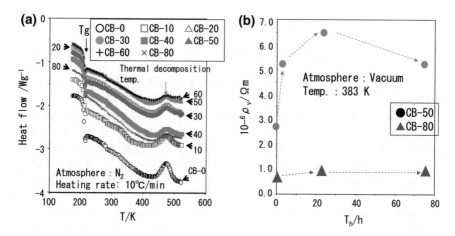

Fig. 3.5 Dependences **a** of heat flow under nitrogen on temperature, T by DSC, and **b** of volume resistivity, ρ_v on heat treatment temperature, T_h in vacuo. Samples; CB-filled NR vulcanizates (CB-50 and CB-80) (from Figs. 2(a) and 5(b) in Ref. [52])

Table 3.2 Characteristics of CB network structure obtained from 3D-TEM image analyses of CB-50 and CB-80: Change of CB network structure before and after heat treatment at 383 K for 75 h in nitrogen gas atmosphere

Samples		L_{cross}^a/nm	L_{branch}^b/nm	D_{cross}^c/nm^{-3}	D_{branch}^d/nm^{-3}	F_{cross}^e	F_{branch}^f
CB-50	Original	40	82	8.0×10^{-6}	6.7×10^{-7}	0.94	0.06
	Heat-treated	23	42	2.0×10^{-5}	1.1×10^{-5}	0.68	0.32
CB-80	Original	39	110	1.7×10^{-5}	3.1×10^{-6}	0.88	0.12
	Heat-treated	22	46	3.0×10^{-5}	1.7×10^{-5}	0.67	0.33

From Table 3 in Ref. [52]
[a,b]Average lengths (L_{cross}, L_{branch}) of cross-linked chain and branched chain of CB network, respectively
[c,d]Densities (D_{cross}, D_{branch}) of cross-linked points and branched points of CB network, respectively
[e,f]Fractions (F_{cross}, F_{branch}) of cross-linked chains and branched chains, respectively

branched chains were evaluated from the 3D-TEM images. Here, these four parameters were defined as follows:

$$D_{cross} = N.Nd/TV, \tag{3.15}$$

$$D_{branch} = N.Tm/TV, \tag{3.16}$$

where TV is the volume observed by 3D-TEM, N.Nd is the number of cross-linked points, and N.Tm is the number of branched points.

$$F_{cross} = N.\text{NdNd}/(N.\text{NdNd} + N.\text{NdTm}), \qquad (3.17)$$

$$F_{branch} = N.\text{NdTm}/(N.\text{NdNd} + N.\text{NdTm}), \qquad (3.18)$$

where $N.\text{NdNd}$ is the number of cross-linked chains, and $N.\text{NdTm}$ is the number of branched chains.

Table 3.2 suggests the changes of CB network structure by heat treatment: L_{cross} and L_{branch} decreased much, and those of two samples were almost same after heating. Further, D_{cross} increased slightly and approached to the almost constant value, too. While D_{branch} of CB-50 much increased, change of CB-80 was minor by heating. These results suggest that the rearrangement and the scission of the CB network in CB-50 were more frequent than in CB-80. We found that after the heat treatment, F_{cross} of the two samples was decreased and F_{branch} was increased, that is, both F_{cross} and F_{branch} of the CB networks were influenced by the heat treatment. In particular, increase of F_{branch} of CB-50 was larger than that of CB-80, while decrease of F_{cross} of CB-50 and CB-80 was not much. Furthermore, F_{cross} and F_{branch} of the heat-treated CB-50 and CB-80 were not much changed, and their values were approximately 0.7 and 0.3, respectively. These results suggest that CB network of CB-80 has more thermal stable than that of CB-50. The CB network structure in non-conductive rubbery matrix affords effective electron conductive paths, contrary to the rubbery matrix itself which is an insulator. While the conductive paths of CB-50 are remarkably cut off by heating, the paths of CB-80 are not damaged much under heat treatment. We believe that these facts are reflected in the behavior of the ρ_v dependence (Fig. 3.5b) on T_h for both samples.

3.2.3 Electrical Property

(1) Rubber as an Insulator, a Dielectric, and an Electron Conductor

Polymer materials including rubber generally show very low electron conductivity and are employed as an electrical insulator. On this point, metals which are highly conductive are in contrast to polymers. However, at the "Electronics Age," conductive polymers have been studied widely since the 1960s. H. Shirakawa who won Nobel Prize in Chemistry in 2000 is a synthetic polymer chemist who found the relatively high conductivity of poly(acetylene) by exposing to dopant like iodine, and further soft polymeric conductive materials are worldwide under development up to now.

Once upon a time, NR was employed as the covering insulator of electrical wires or cables due to its flexibility. Flexibility originated from rubber elasticity is the characteristics when we consider using rubber for electronics. For example, the conductive soft composites by mixing rubber with conductive fillers, such as carbon black (CB), have been investigated for long, and lots of soft devises and articles are on the market so far. Refer to the books [56, 57] for the relevant information to these composite materials.

The insulating materials including rubber are not conducting, but are dielectric from the viewpoint of electromagnetism in modern physics. Namely, the polarity of insulating rubber is low, but the rubber possesses electrical dipole moment. Though an electric current does not flow in rubber even placed under an electric field, polarization occurs through the orientation of dipoles. This is a dielectric, and it functions as a condenser of electronic charge by polarization under DC field to make a storage capacitor. The response of dielectrics under AC electric field is reflected to the physicochemical behaviors, which are assumed to be closely related to the molecular-level dynamics of dipoles in them. In other words, observed behaviors of them under AC field give data of value for a study of structure and physical properties of polymer materials through the motion of dipoles [58, 59]. Recent studies on electric conductivity and dielectric characteristics of CB-loaded rubber vulcanizates are explained below.

Some rubber matrices may afford excellent lithium-ion conduction ability, which is of much use for lithium-ion batteries. For this ionic application of rubbery materials, see Sect. 3.3.3.

(2) **Volume Resistivity of Nanofiller-Loaded Rubber**

Figure 3.6 shows the volume resistivity (ρ_v) of CB-loaded NR vulcanizates as a function of the amount of CB loading at various temperatures between room temperature (296 K, or 23 °C) and 363 K (90 °C) [51]. At the CB loading up to 10 phr, the change of ρ_v was not distinguished, while ρ_v decreased drastically at the CB loading higher than 10 phr down to almost one-third of the initial values at about 40 phr loading. The observed phenomenon during which the samples have changed from an insulator to a semiconductor in a sudden is called electrical percolation. The threshold of this sudden change is approximately at 35 phr of CB loading. It is said that a conductive integrated circuit or an electric network is completed when the value of ρ_v attains constancy [60]. This completion seems to be at 50 phr CB loading as shown in Fig. 3.6.

On the other hand, Blythe suggested that the hopping-conductive mechanism, by which an electron gets over a barrier, and the tunnel effect, by which an electron goes through a barrier, can be distinguished [59]. We found in Fig. 3.6 that ρ_v decreased with increasing temperature at any CB loadings. This fact suggests that conductive mechanism in CB-loaded NR vulcanizates is a thermal activation process. Figure 3.7a shows the Arrhenius plots of conductivity (σ) which is the inverse of ρ_v, for different CB loading levels. From the approximated slopes, activation energies for conduction are calculated. As shown in Fig. 3.7b, the activation energy ΔE decreased dramatically with CB loading, and the ΔEs are all nearly equal after 20 phr loadings. While very large activation energy values ($\Delta E_\sigma = 90 \sim 120$ kJ/mol) were observed below 20 phr of CB, approximately its one-fifth values, $\Delta E_\sigma = 20 \sim 40$ kJ/mol, were observed at higher CB loadings. The latter values are nearly

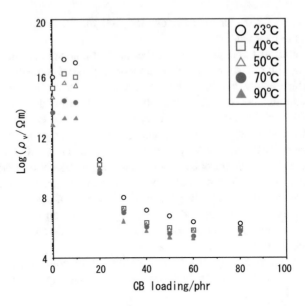

Fig. 3.6 Relationship between volume resistivity (ρ_v) and CB loading of CB-filled NR vulcanizate in temperature region from 23 to 90 °C (from Fig. 6 in Ref. [51])

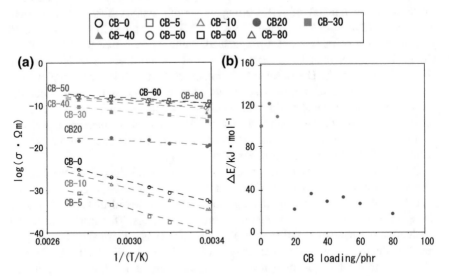

Fig. 3.7 **a** Arrhenius plots of conductivity σ (= $1/\rho_v$, ρ_v volume resistivity) and **b** activation energy ΔE_σ of CB-filled NR vulcanizates (from Figs. 7 and 8 of Ref. [51])

equal to those which were reported by Jaward et al. [61] for the activation energy found for the rubber compounds with 45 and 60 phr of graphitized carbon.

This suggests that an electron path in the network structure is formed in the high CB loading region, and electrons move through the path formed by CB networks with the relatively low activation energy by electron hopping and/or the tunnel effect. Which mechanism, hopping or quantum mechanical tunneling, is still a question to solve. In addition, the observed conductivity at lower CB loadings, which is much smaller, may still suggest a little electron conduction through the rubbery matrix supposedly due to some semiconducting impurities. Namely, electron hopping via non-rubber additives such as curing reagents might have contributed to the conduction. In order to reduce it further, whether extracting the impurities from rubber vulcanizates may be recommended, or usage of peroxide-cross-linking rubber may be recommendatable, instead of sulfur vulcanization.

(3) Dielectric Relaxation Behaviors

The dielectric relaxation behaviors of polymer under AC field are known to be influenced by quite a lot of factors: chemical structure of repeating units in the polymer chain, size and characteristics of segments, constraint of segmental motions by the higher-order structure, existence of electronic charges and their distribution if any, configuration of permanent dipoles in the chains, anisotropy of the polarizability, and so on. Note that many of these are molecular-level factors, and hence, elucidation of dielectric behaviors may afford molecular-level structural information of much value.

In the analysis of dielectric relaxation results, interpretation of the so-called Cole-Cole plot is of importance. The vertical and the transverse axes of the plot are ε'' and ε', respectively. Here, the complex dielectric constant ε^* is defined as Eq. (3.19).

The Debye-type single relaxation is described by the following equations [58, 59, 62]:

$$
\begin{aligned}
\varepsilon^* &= \varepsilon' - i\varepsilon'' \\
&= \varepsilon_\infty + (\varepsilon_S - \varepsilon_\infty)/(1 + i\varpi t),
\end{aligned}
\tag{3.19}
$$

where ε' and ε'' are permittivity and dielectric loss of complex permittivity, respectively. ε_∞ is instantaneous permittivity which is measured just after an electric field is applied, and ε_S is static permittivity which is measured after a sufficient period of time after the application of electric field. Further, i, ω, and t are imaginary unit, angular frequency, and time, respectively. From Eq. (3.19), the following equations are obtained.

$$
\varepsilon' - \varepsilon_\infty = (\varepsilon_S - \varepsilon_\infty)/\left[1 + (1 + i\varpi\tau)^2\right],
\tag{3.20}
$$

$$
\varepsilon'' = (\varepsilon_S - \varepsilon_\infty)\omega\tau/\left[1 + (\omega\tau)^2\right],
\tag{3.21}
$$

where τ is dielectric relaxation time which is the time needed to return to the $1/e$ of the random equilibrium state, namely, τ is the index of mobility of dipole moment.

When Eqs. (3.20) and (3.21) are combined by eliminating $\omega\tau$, Eq. (3.22) is obtained.

$$[\varepsilon' - (1/2)(\varepsilon_S + \varepsilon_\infty)]^2 + (\varepsilon'')^2 = [(1/2)(\varepsilon_S - \varepsilon_\infty)]^2. \qquad (3.22)$$

This equation shows that the Cole-Cole plot of the Debye-type single relaxation shows a semicircle. The dielectric characteristics (permittivity ε' and dielectric loss ε'') of the CB-filled NR vulcanizates were investigated at room temperature in a frequency range of 20 Hz to 1 MHz [51, 52, 63, 64]. One result is shown in Fig. 3.8 [52]. Here, any notable patterns are not observed in the CB-loaded samples of lower than 10-phr CB, and only an incomplete pattern of a circular arc is observed in 20-phr sample. Therefore, they are not included in the figure. In contrast, the NR samples of loaded amount of CB being higher than 30 phr clearly show a circular arc-shaped pattern at the high-frequency region.

Moreover, this arc-shaped pattern becomes larger, when the CB loading is higher. This arc-shaped pattern may be attributable to some type of relaxation phenomena, it is inferred that the CB in NR has aggregated to form a particular structure, namely CB networks at the CB loading higher than 30 phr. And also, the network structure grows with the increase of CB loading. The value of 30 phr CB loading correlates well with the CB loading threshold mentioned earlier for the formation of structure. The CB-20 sample displayed only a portion of the circular arc-shaped pattern, i.e., the pattern was incomplete compared with those seen in the samples of CB loadings higher than 30 phr. Therefore, the CB-20 sample was excluded from the subsequent analysis described below. By the way, the relaxation of the normal mode, which is related to fluctuation of the vector between the ends of

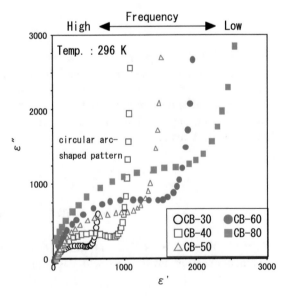

Fig. 3.8 Cole-Cole plots for CB-filled NR vulcanizates showing dielectric loss (ε'') against permittivity (ε') of CB-30, CB-40, CB-50, CB-60, and CB-80 (from Fig. 15(b) in Ref. [51])

a molecular chain of polyisoprene, is observed in the frequency region between 20 Hz and 1 MHz [65].

Next, procedures of an original graphic analysis method applied to the Cole-Cole plots are briefly described here [60, 61]. In Fig. 3.8, an empirical equation is used to remove the relaxation component on the low-frequency side from the circular arc-shaped pattern, which is observed in the CB loading higher than 30 phr. Then, a geometrical analysis is applied in order to separate the semicircular relaxation component and the remaining residual component [60, 61]. These two relaxation components (semicircular and residual components) are shown in Fig. 3.9. In the figure, the vertical axis shows $\Delta\varepsilon''$ which represents the remaining value after removing the low-frequency relaxation component from the measured ε'' value. For the samples with CB loading of 30 phr or higher, it is possible to separate the two relaxation components in this way. The results indicate that the both relaxation components increase with higher CB loading and they also tend to extend into the high ε' region (low-frequency region). It is noted that virtually no relaxation components were observed for the samples with CB loading of 20 phr or less.

ε'_{∞} and ε'_s are the ε' values when the frequency is infinite and when the frequency is 0, respectively. The relaxation strengths of the semicircular component ($\Delta\varepsilon'_{cir}$) and the residual component ($\Delta\varepsilon'_{res}$) may be defined with respect to the difference between ε'_{∞} and ε'_s. Furthermore, the fractions (F_{cir} or F_{res}) of the relaxation strength of each component are obtained from the results of 3D-transmission electron microscope (3D-TEM) observation and image analysis which will be described in Sect. 4.1.3. Thus, the fractions (F_{cir} and F_{res}) of the relaxation strength of each component can be expressed as

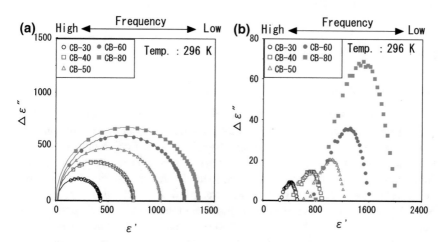

Fig. 3.9 Two relaxation components of CB-30, CB-40, CB-50, CB-60, and CB-80: **a** semicircular and **b** residual relaxation components. $\Delta\varepsilon''$ is defined as the value obtained by subtracting an extrapolated curve of a Cole-Cole plot on the high-frequency side from the circular arc-shaped relaxation. The distances between two points where the curves of the two relaxation ($\Delta\varepsilon'_{cir}$ and $\Delta\varepsilon'_{res}$) cross the ε' axis, are the relaxation strengths of the two relaxation components (from Fig. 19 in Ref. [51])

$$F_{\mathrm{cir}} = \Delta\varepsilon'_{\mathrm{cir}} / \left(\Delta\varepsilon + \Delta\varepsilon'_{\mathrm{res}}\right) \text{ and } F_{\mathrm{res}} = \Delta\varepsilon'_{\mathrm{res}} / \left(\Delta\varepsilon'_{\mathrm{cir}} + \Delta\varepsilon'_{\mathrm{res}}\right).$$

In accord with the results in Refs. [51, 63, 64], a good positive linear correlation is seen between the fraction of relaxation strength of the semicircular component and F_{cross} and between the fraction of relaxation strength of the residual component and F_{branch}. These results suggest that the CB aggregate network structure is closely related to the dielectric relaxation characteristics. In other words, the semicircular relaxation component and the residual relaxation component of dielectric relaxation are probably attributable to the polarization of the rubber around the cross-linked chains and to the interaction between the branched chains and the CB aggregate network.

On the other hand, as we see later in Sect. 4.1.4, we estimate that there are a few interaction layers that their mobility is restricted in the vicinity of CB, and CB network is formed by connection of CB aggregates with the layer the mobility of which is most restricted within them. This most restricted rubber which may or may not include cross-linked chains of CB networks should be less free than the branching chains. Whereas the relaxation component of the strongly restrictive layer is observed in the high-frequency region in the Cole-Cole plot, that of the weakly restricted layer is observed in the low-frequency region. This means that the Cole-Cole plot of CB-loaded rubber vulcanizates displays a deformed circular arc-shaped pattern, presumably due to the network structuring of CB. Therefore, we assume that the semicircular relaxation component in Fig. 3.9a and the residual one in Fig. 3.9b are closely related to the highly restricted rubber layer and the less restricted one, respectively.

3.2.4 Optical Property

(1) Rubber and Optics?

Generally speaking, rubber is not excellent in transparency, and there have not been many examples in which the optical property of rubber is an important issue. The transparent rubber in the past was so-called "amber-color rubber", and was used for rubber bands, rubber threads, slender rubber tubes, rubber nipples for children and heels for shoes, and so on. One of a few cases for which the transparent rubber was industrially requested was from the electronics industry, which had a demand for a good encapsulating material of the electric circuit, both optically transparent and electrically insulating. Rubber was an insulating material except for an electroconductive rubber (see previous section), and the rubber which could respond to the above demand was only silicone rubber from the viewpoint of transparency [66, 67].

A more recent demand may be a flexible optic fiber of transparent silicone rubber that enables us to install an optical communication network at places with a lot of corners in the interior of building [67]. Even though a big feature of silicone rubber is a high chemical stability, some chemical modifications are possible. Among them, chemical reaction of hydrosilyl group (–Si–H) with vinyl compounds (CH_2=CH–) is well known as hydrosilation [67–70]. When a mesogenic compound carrying a vinyl group is subjected to the reaction with silicone rubber containing silanol groups, liquid crystalline elastomer (LCE) is easily synthesized [71]. The liquid crystal [71–73] has multidimensional demand for electrooptic devices as a display panel and so on. These optical applications may be a challenging field for rubber in a near future.

When silicones are excluded, transparent elastomers are limited to some experimentally synthesized ones in research laboratories: The elastomer, which consists of oxytetramethylene linkage and viologen structure, is of ionene type with nitrogen dication (N^+–N^+) in the main chain. It displayed an excellent transparency as well as good mechanical properties [74–76]. The ionene elastomers in general are one kind of TPE explained in Sect. 3.5.3 in which cations (N^+) associate into an ionic aggregate to result in network-like structure formation. It is estimated that the ion aggregates function as a cross-linked point. Additionally, the dication in the viologen unit is reduced to a cation radical by one-electron reduction under photo-irradiation. The reduction is accompanied by a color change from colorless to green, reversibly. This color change by photon is called photochromism, and it is one of the most interesting photoreaction of elastomers. When the dication radical in main chain is reduced to the cation radical, the mechanical property also changes in accordance with the change of the association state by the reduction of a dication to a cation radical. Therefore, this elastomer exhibits a photomechanical behavior, too.

From the viewpoint of rubber elasticity, polymer gel is an important substance closely related to rubber among soft materials. A highly motivated study on liquid crystal gels has started by using low-molecular liquid crystal as a swelling solvent of cross-linked butadiene rubber (BR) [77], followed by basic studies on liquid crystalline gels [78, 79]. Now, the liquid crystalline gels have attracted great attention from the application arena as new stimulation-responsive soft material [80, 81]. Comparing with the other science field, the well-planned research works which are fundamentally conducted on the basis of rubber elasticity might more unaffectedly be developed into the application stage, even without an extraordinary attention onto applying. Rubber elasticity is so unique a property that it may be easier to find some applications (see Sect. 3.3.1, too).

The automotive tire is always associated with "black" color, because CB as reinforcing filler is mixed with rubber. However, in the case of rubber to which white or transparent reinforcing filler like silica is added, there is a possibility of non-black-colored tires. Mixing of a pigment onto silica-loaded rubber tire affords red, yellow, or any colored tires. Recently, the "green" tire [82], prepared by the rubber compounded with silane coupling agent as well as silica [83, 84], was commercialized. Here, green does not mean color, but it means ecologically friendly or more sustainable than the CB-filled tire. Possibly, it will be also pioneering the way to multicolored automobile tires. Moreover, the unprecedented silica compounding method like *in situ* silica generation in the rubbery matrix (see Sect. 2.2.3) has been developed, and the use of non-black reinforcing filler is spreading (see Sect. 5.4.3). Whereas CB is mass-produced from crude petroleum, the use of silica is an expression of the recent trend, the so-called "decarbonization trends".

This section introduces a few recent studies of rubber in the optical field. Optics has been a blank arena of rubber except the issue in which the light resistance is necessary for rubber products. Thus, there are not a lot of rubber examples in this field yet. Acceleration of the developmental study on the optics of rubber is advisable for our future, from the viewpoint of sustainable development.

(2) Refractive Index of Silicone Rubber

For the purpose of controlling the refractive index of an optically transparent silicone rubber, various substituent groups were introduced into the rubber using hydrosilation reaction (see Fig. 3.10). This investigation may be just a beginning of rubber in optics, suggesting the possibility of manufacturing rubbers of various refractive indices. It is well known [85] that by introduction of aromatic (rich in π-electron) substituent groups into vinyl polymer, the electric polarization and the refractive index increase. In contrast, when a substituent group including fluorine (high in electron negativity) is introduced, its refractive index becomes small. By using 4 kinds of silicone rubbers having the different contents of hydrosilyl group as a reactant, the introduction of various chemical groups by hydrosilation reaction was conducted as shown in Fig. 3.10 [67].

Figure 3.11 shows the obtained results, indicating the relationship between the refractive index (n_D^{20} is measured at 20 °C using the sodium D-lines.) and the Abbe number (which is also called the reciprocal dispersion power and indicates the dependence of refractive index on the wavelength of light) of the product. In this figure, the other polymers as well as inorganic glass were shown by small circles with number and English abbreviations, respectively. The refractive indices of the chemically modified poly(dimethylsiloxane)s are in the range from 1.36 to 1.69. The values are widely changed around the refractive index of poly(dimethylsiloxane) (small circle numbered 17 in Fig. 3.11). The Abbe number changes from 80 to 20 and decreases with increasing the refractive index. This change of the Abbe number suggests that the higher is the refractive index, the higher is the dispersion. In other words, it is difficult to get high refractive index silicone elastomer with low wavelength dependency. This tendency is not favorable for using as an optical device. The chemically modified poly(dimethylsiloxane) cannot avoid this tendency as long as the refractive index depends on high electronic polarizability [85]. Therefore, molecular design and synthesis of polysiloxane which has the big Abbe number (namely the smaller dependence on wavelength) as well as the high refractive index, would be the future issue to be investigated.

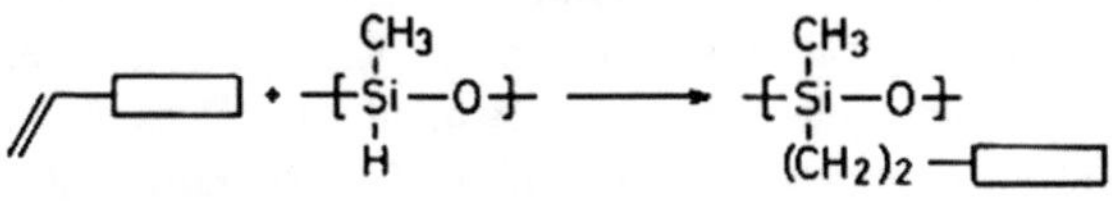

Fig. 3.10 Hydrosilation reaction of vinyl compounds (from Fig. 2 in Ref. [67])

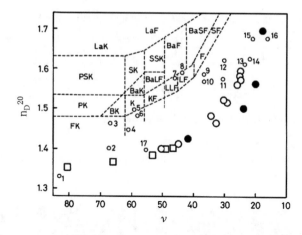

Fig. 3.11 Refractive index (n_D^{20}) and Abbe number (ν) of chemically modified silicone rubber. In the figure, *square, white circle,* and *black circle* designate the silicone rubbers with the introduced fluorine group, naphthalene group, and anthracene group, respectively. *Small open circles* with the number are the data of commercially available polymers: No. 17 is that of commercial silicone elastomer (from Fig. 13 in Ref. [67])

(3) **Optical Transparency of Silica-Loaded Natural Rubber**

Silica as another nanofiller for tire is recently focused (see Sect. 5.4.3). Here, a study on optical transparency of particulate silica-loaded NR cross-linked by peroxide is explained [86–88]. The recipes for the hydrophobic (VN-3) and hydrophilic (RX 200) silica-loaded cross-linked NRs examined in this study are shown in Tables 3.3 and 3.4, respectively. The amount of dicumyl peroxide (DCP) was set at 1.0 phr (parts per hundred rubber in weight). Only the silica loading was changed from 0 to 80 phr in the cross-linked NR. After kneading rubber compounds using an open two-roll mill, the cross-linked rubber sheet of 1 mm thickness was prepared by compression molding at 155 °C for 30 min. The hydrophobic silica-filled cross-linked NR is designated as RX# sample, and the hydrophilic silica-filled cross-linked NR as VN# sample. Here, # indicates the amount of silica loading in phr.

Table 3.3 Compounding recipes for cross-linking of hydrophilic silica-loaded NR[a]

Sample	VN0	VN10	VN20	VN30	VN40	VN60	VN80
NR	100	100	100	100	100	100	100
DCP[b] (phr[c])	1	1	1	1	1	1	1
Silica VN3[d] (phr)	0	10	20	30	40	60	80

From Table 7.1 in Ref. [87]
[a]Cross-linking conditions: 30 min at 155 °C under 100-150 kg/cm^2
[b]Dicumyl peroxide
[c]Part per one hundred rubber by weight
[d]Nipsil VN-3 (average primary diameter = *ca.*16 nm) from Tosoh Silica Corporation

Table 3.4 Compounding recipes for cross-linking of hydrophobic silica-loaded NR[a]

Sample	RX0	RX10	RX20	RX30	RX40	RX60	RX80
NR	100	100	100	100	100	100	100
DCP[b] (phr[c])	1	1	1	1	1	1	1
Silica RX[d] (phr)	0	10	20	30	40	60	80

From Table 7.1 in Ref. [87]
[a]Cross-linking conditions: 30 min at 155 °C under 100–150 kg/cm^2
[b]Dicumyl peroxide
[c]Part per one hundred rubber by weight
[d]AEROSIL RX200 (silica modified by trimethylsilyl group, average radius = *ca*. 12 nm) from Evonik Degussa Japan Co., Ltd

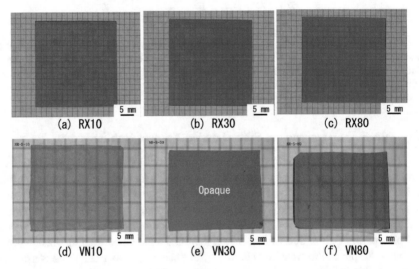

Fig. 3.12 Optical transparency of hydrophobic (RX10, RX30, and RX80) and hydrophilic (VN10, VN30, and VN80) silica-loaded NRs (from Figs. 7.2 and 7.3 in Ref. [87])

Figure 3.12 shows images of the optical transparency of the RX-series and VN-series NR samples obtained by placing them on a white section paper [87]. The images in Fig. 3.12a–c show that the rubber sheets look like the same in spite of increasing quantity of the hydrophobic silica filler. Whereas the optical transparency is usually assumed to decrease by mixing a foreign body, not much change in transparency even by increasing the loading of hydrophobic silica filler is somewhat extraordinary. In contrast, the images in Fig. 3.12d–f show that the visibility of graph paper behind the rubber sheet decreased in the order of VN10 ≥ VN80 >> VN30. At the silica content of 30 phr, the section paper was invisible. When the silica loading was further increased to 80 phr, the pattern of the graph paper became visible again. It is reported that the increase of silica or glass fiber compounded into transparent polymers generally tends to reduce the optical transparency of the composite materials [89]. The reason of observed maximum opacity at 30 phr hydrophilic silica content has to be specifically explained, and further study on a few optical properties relating to transparency is undertaken on the two silica-filled samples as follows.

Figure 3.13 illustrates the concepts of diffusion transmittance and haze, both of which are used effectively in optics-related industries as indices of transparency in the visible light region [85, 86]. Excluding the scattering of light caused by irregularities on the sample surface, two cases may be considered when a sample is opaque. In one case, incident light is absorbed or scattered inside the material, resulting in a shield effect that blocks the transmission of light. In the other case, light transmitted from the material is markedly diffused or scattered. Accordingly, the shield effect can be quantitatively examined in the former case, while diffusion transmittance and haze can be similarly examined in the latter case. The shield effect (T_{shield}), diffusion transmittance (T_d), and haze (H) are defined as the following Eqs. (3.23–3.25), using total transmittance (T_t) and parallel transmittance (T_p) [90]. It is noted that all these parameters (T_{shield}, T_t, T_d, and T_p) are expressed in percentage, and in the case that the values of T_{shield}, T_d and H are lower, the transparency of the material is higher

$$T_{shield} = 100 - T_t, \tag{3.23}$$

$$T_d = T_t - T_p, \tag{3.24}$$

$$H = 100(T_d/T_t), \tag{3.25}$$

where the parameters (T_{shield}, T_t, T_d, T_p, and H) are the values from 0 to 100%.

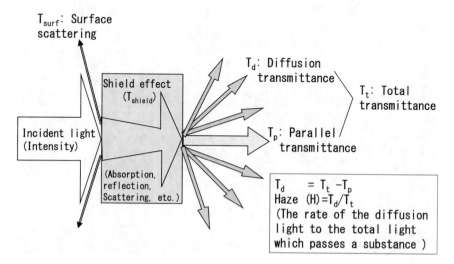

Fig. 3.13 Definition of shield effect (T_{shield}), diffusion transmittance (T_d), parallel transmittance (T_p), and haze (H) (from Fig. 1 in Ref. [86])

On the other hand, the formation of silica network structure originated from silica aggregation was observed using the three-dimensional transmission electron microscopy (3D-TEM), which is explained later in Sects. 4.1.2 and 4.1.3. In the high silica loading region where d_p (distance between silica aggregates) is shorter, silica aggregates further coalesce to form an agglomerate, ultimately leading to a network structure of particulate silica. Consequently, d_p saturated at the loading of 40 phr and almost a constant d_p value of ca. 1.3 nm was observed regardless of the silica surface being hydrophobic or hydrophilic. The value of 1.3 nm suggests that there is a rubber layer, i.e., bound rubber layer around the silica aggregates that prevents them from directly contacting each other. It is inferred that the network-like structure of silica aggregates, i.e., a kind of agglomerate formed via this bound rubber.

The value of d_p = 1.3 nm was used here to visualize the structure of the silica aggregates, in particular, the network structure at high silica loadings. Specifically, the constant distance 1.3 nm is interpreted as the two aggregates are virtually in contact. By connection, the neighboring aggregates of 1.3 nm distance are connected to form diagrams as shown in Fig. 3.14. Here, the connecting by line between the centers of gravity of nearest-neighbor (1.3 nm distance) silica aggregates is conducted on the 3D-TEM image, to result in a diagram shown in the figure. Figure 3.14 therefore shows the network-like structures of hydrophobic and hydrophilic silica aggregates in cross-linked NR samples for silica filler loadings of

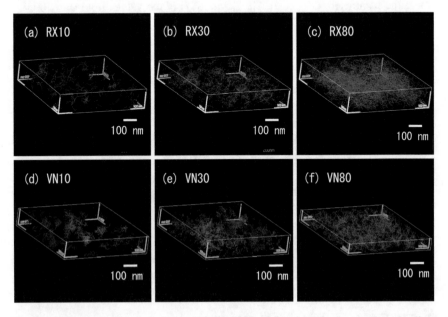

Fig. 3.14 Silica network in cross-linked NR. See Tables 3.3 and 3.4 for the sample codes (from Fig. 9 in Ref. [88])

10, 30, and 80 phr. Imperfect networks that are not completely connected (the so-called pregel states) are shown in different colors in the figure.

While the hydrophobic silica aggregates are nearly all connected to result in a network structure at loadings of 30 phr or higher, the hydrophilic silica aggregates show less tendency to perfect networks. This is especially recognized by comparing two diagrams at 30 phr silica loading samples. In the case of hydrophilic silica, locally unconnected aggregates to the network are seen here and there, and the perfect connection among all the hydrophilic silica aggregates is seen only at the loading of 80 phr. These results are in accord with a conventional empirical information known among rubber technologists, that is, the effect of stronger filler-to-filler (relative to filler-to-rubber) interaction in hydrophilic silica than hydrophobic silica and carbon black is an important factor to be considered for utilization in rubber [91, 92].

A more detailed observation of Fig. 3.14 reveals that the network structure is consisting of three kinds of chains, that is, cross-linked chains (cross-linking points), branched chains (branching points), and isolated chains that are not connected to the network. It is assumed that the cross-linked, the branched, and the isolated chains in the silica networks are the light scattering media under irradiation of light [88]. Employing a multiple scattering equation for a scattering medium derived by the Monte Carlo method, which is reported by Kandidov et al. [93], the relationship is investigated between the haze and the product of the number density of the chains and the square of their average length. The equations are shown below:

$$H \sim \mu_s (= \pi K_p n_{\mathrm{iso}} r_{\mathrm{iso}}^2) \sim (\pi K_p/4) n_{\mathrm{iso}} L_{\mathrm{iso}}, \tag{3.26}$$

$$H = C n_{\mathrm{iso}} L_{\mathrm{iso}}^2 + D(H), \tag{3.27}$$

$$L_{\mathrm{iso}} = 2 r_{\mathrm{iso}}, \tag{3.28}$$

$$C = \pi K_p/4, \tag{3.29}$$

where H, μ_s, and π denote haze, the scattering coefficient (m^{-1}), and the circular constant, respectively. The n_{iso}, r_{iso}, and $D(H)$ indicate the number density of the isolated chains (m^{-3}), one-half of the mean length of the isolated chains (L_{iso}), and an additional constant on the cross-linked rubber, respectively. It was found that the isolated chains in the hydrophilic silica networks are responsible for the optical transparency to decline during the multiple scattering of light. Product of the number density and square of the length ($n_{\mathrm{iso}} L_{\mathrm{iso}}^2$) of the isolated chains in the silica networks depended on the hydrophobic and hydrophilic silica loading; $n_{\mathrm{iso}} L_{\mathrm{iso}}^2$ decreased with increasing silica loading in the case of hydrophobic silica, $n_{\mathrm{iso}} L_{\mathrm{iso}}^2$ showed the maximum value near 30 phr in the case of hydrophilic silica.

Fig. 3.15 Dependence of
haze (*H*) on $n_{iso}L_{iso}^2$ for
hydrophobic and hydrophilic
silica-loaded NR (from
Fig. 15 in Ref. [87])

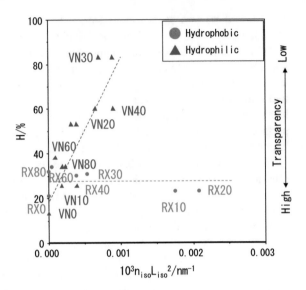

Figure 3.15 shows the dependence of haze (*H*) on n_{iso} L_{iso}^2 for the hydrophobic
and hydrophilic silica-filled cross-linked NR in accordance with Eqs. (3.26) and/or
(3.27). In the case of hydrophobic silica, *H* did not display any dependence on n_{iso}
L_{iso}^2. On the other hand, hydrophilic silica showed a linear dependence of *H* on n_{iso}
L_{iso}^2, as indicated in Eq. (3.27). Simultaneously, the decrease of the transparency
observed in Fig. 3.12 was well reproduced in the order of

$$VN0 > VN10 > VN60 \fallingdotseq VN80 > VN20 \fallingdotseq VN40 > VN30.$$

These results suggest that the multiple scattering of light by the isolated chains in
the silica networks has caused optical transparency to decline in the hydrophilic
silica-filled cross-linked NR samples.

From the above mentioned, in the hydrophilic silica-loaded NR, the matrix
rubber governs the optical characteristics at silica loadings lower than 30 phr, since
a small amount of isolated chains are formed in the system. While the optical
characteristics due to the cross-linked and branched network chains may uniform
even in the thickness direction of the specimens, the isolated chains may probably
act as non-uniformly distributed light scattering bodies. In addition, it is inferred
that the optical characteristics of the network chains are dominant at filler loadings
higher than 30 phr, because isolated chains are markedly scarce. Still, the exact
reason why isolated chains noticeably increase at a hydrophilic silica loading of 30
phr is not clear. It is hypothesized that this increase in isolated chains may be
dependent on strong filler-to-filler interaction in hydrophilic silica and the
cross-linking conditions using peroxide as a cross-linker, and it is necessary to
conduct more detailed investigation on these points in a future work.

3.3 Development of Highly Functional Elastomeric Devices

3.3.1 High Functionality of Rubber and Elastomer

Rubber and elastomer have been utilized as materials to prepare various devices for our comfortable living and safe life in our society. In fact, a rubber bowl prepared from NR had been used on the religious or political occasions for rites and festivals in the Olmec and the following civilizations until the Spanish invasion in the sixteenth century. There must have been manufactured some more kinds of rubber goods in the long history of people, especially in the Central and South America. However, the first rubber good leading to the modern use of NR is a raincoat: An English manufacturer S. Peal patented a waterproofing cloth by applying a solution of NR in turpentine in 1791. In 1823, C. Macintosh (1766–1843) manufactured double-textured rainproof coats and started selling in England. The two clothes on which NR solution in naphtha oil were applied and were laminated so that the NR-coated layers are put between the two clothes in order to avoid the rubber layer to be surfaced. Thus, obtained coats were called "mackintoshes", and the name remains up to now for raincoats [94]. Two early examples suggest one function of rubber, which is its repellent and non-permeating ability of water or moisture.

The tire is one of the structural materials supporting the whole automobile (itself and the driver, some passengers, and perhaps luggage and freight) on the road surface, i.e., a structural material. However, much more important is its functions enabling a safe and comfortable drive (see more in Chap. 5). The dynamic contact of tire tread (made of vulcanized rubber) with the road surface ensures the safe and stable driving, a marvelous functionality device showing friction, rolling resistance, grip on the road surface, and wear to minimize the wear of the road. Without rubber elasticity, rubber seals and rubber belts fail to play a role for sealing and to transmit the movement smoothly, respectively. In addition, rubber in general and rubber tires in particular have been absolutely necessary items in our modern society, which is sadly but clearly suggested during the two World Wars in the twentieth century, when the rubber was indispensable as a strategic material [94]. These examples suggest that rubber elasticity is not only a simple physical property but also a high functionality itself, highly useful to the wide range of performance.

Most of the rubber products are thus well known to show various useful functions in addition to the rubber elasticity, which is intrinsic to rubber. Rubber elasticity may be better classified as one original function by itself. In this section, however, some special functions other than rubber elasticity are to be discussed. Even in this section, the mechanical properties are also mentioned, since almost all the functions of rubber are under strong influence of its cross-linking and network structures, which are the determining factors of mechanical properties, too.

From a viewpoint of functionalization, there are three directions to afford functionality other than rubber elasticity:

1) The first is a utilization of rubbery matrix as a polymeric solvent. Functional fillers, especially particulate ones, are usually mechanically mixed into rubber matrix. The reinforcing fillers such as carbon black (CB) are thus mixed with rubber by mixer, taking advantage of the "polymer solvent" nature of rubber matrix [95, 96], which is quite common in rubber industry. In the case of conductive rubbers, for example, various kinds of conductive particles (including metal powders and a few grades of CB) are mechanically mixed to rubber to afford soft conductive or semiconductive devices, which have been of much use in practical usages. One particular example using this technique, suggests a pressure-sensitive conductive rubber, a kind of transducer, was prepared by controlling the morphology of electroconductive fillers in the rubber matrix as shown in Fig. 3.16. Namely, a pressure sensor is produced by mechanical mixing of the filler and the rubber: The electron-conductive path is not connected from the top to the bottom under a pressure F = 0 as shown in the left-hand side in the figure. However, by increasing the pressure the number of the connected path is increased; hence, the electric current is passable. That is, soft and flexible pressure sensor is obtained [95, 97].

This type of soft transducer is of much value, for example, for electrodes of a biomedical instrumentation, which are applied on the skin of patients. Flexibility of the transducer enables a very good contact between the electronic device and skin. Note that the preparation of this kind of sensor by filler loading onto rubber is much more difficult and complicated than just mixing: Excellent techniques for controlling the dispersion as shown in Fig. 3.16 are mandatory for giving a function of sensor, and it was finally achieved probably after lots of trial-and-error experiment by some experts, who have been engaging in rubber mixing for many years. The sensor is further requested to have a good reproducibility, for which the change of morphology of filler dispersion in the rubber matrix should be completely reversible even after thousands of loading and unloading cycles. Paradoxically, the filler loading onto rubber by mechanical mixing is apparently very simple and easy, but there still remains much room for further advancement.

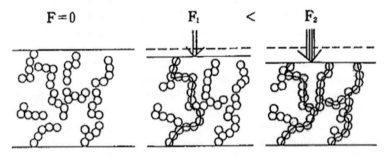

Fig. 3.16 Pressure-sensitive rubber by controlling the morphology of electro-conductive fillers in rubber matrix. F is an external stress, pressure. (From Fig. 12 in Ref. [97])

2) The second is a grafting of functional molecules or groups onto rubber chains by using polymer reactions. For example, the chemical modification of silicone rubber was reported [67] to prepare the functional rubbery materials of various refractive indexes (see Sect. 3.2.4). Generally, a glass transition temperature (T_g) often shifts to the high temperature when a functional group is introduced onto rubber by grafting or by any other polymer reactions, which usually results in the decrease of the elastic nature of rubber. Therefore, it is inevitable for functional rubber materials to have a maximal amount of functional moieties to be introduced by polymer reactions.

3) The third is a utilization of thermoplastic elastomers (TPE, see Sect. 3.5.3). One of the reasons why the cross-linking systems are not preferable for preparing functional materials is a possibility for cross-linking to inhibit or damage the functionalities, and/or one of the curing reagents might be reactive with the functional groups. For example, the sulfur cross-linking systems are usually to be avoided for rubber materials of biomedical and health uses. Lots of curing reagents, organic accelerators in particular, are not compatible with living organs or tissues, i.e., not safe enough for human bodies. When a covalent bonding is necessary for cross-linking in these rubbery materials, a peroxide cross-linking or gelation by condensation-type reactions (e.g., esterification) using three- or tetra-functional monomers is often recommendable, if functional materials are the final target.

In this chapter, a key in the materials design to prepare functional bioactive elastomers and ionic conductive elastomers is focused as two typical examples. The former topic is a study on the biocompatible polyurethaneureas (a kind of TPE) using novel prepolymers which were designed on the basis of the function of living tissues [98–105]. The latter was a study on preparation of ion-conductive elastomers which were designed by taking an advantage of amorphous character of rubber [96, 101, 106–117]. Two approaches may be assumed to be chemically rational, but there already have reported various kinds of functionality rubbers for various applications nowadays. It is hoped to check more examples and a future possibility after reading this section.

3.3.2 Bioactive Elastomers

Many kinds of synthetic biomaterials are now available together with medical instruments for medical diagnosis and treatments. Requested necessary properties of the biomaterial are quite various and depend on the specific purposes and involved parts of the patient's body to which they are applied. Most organs and tissues of living bodies are basically soft and elastic except bones and teeth. Therefore, the materials should be more or less elastic for biomedical uses. When think of a biomaterial for artificial hearts and/or blood vessels, the required properties include not only blood compatibility but also a good mechanical matching with organs or tissues. Both vein and artery are quite soft and elastic, and the mechanical properties of artificial ones should be in accordance with the living ones. One difficulty, in particular, is the excellent mechanical matching at the joint between the artificial and natural vessels. Insufficient matching often results in the hydrodynamic eddy in blood flow, which may accelerate thrombus formation to make the blood flow difficult. Therefore, it is necessary to evaluate the mechanical properties of biomedical materials as well as biocompatibility from a viewpoint of biomechanics [118]. Furthermore, both biological and mechanical evaluations should include in vivo tests as well as in vitro ones for a final clinical judgment [119].

Polyurethane (PU) is a general term for the polymers synthesized from diols, diamines, and diisocyanates by polyaddition reaction. Its elasticity has been recognized from the 1940s, and an elastic fiber "Spandex Fiber" (Lycra®, later Biomer®) was commercialized by Du Pont in 1954 [120]. The reactions between diisocyanate and diol or diamine generate urethane or urea linkages, respectively. The groups with urethane and urea linkages are significantly aggregated by hydrogen bonding, followed by a formation of hard-segment domains. Hard-segment domains are dispersed in the soft-segment matrix to result in the formation of microphase-separated structure (see Sect. 3.5 and Fig. 3.22b). This type of urethane polymers (PUs) is called segmented PU, and those to which amine compounds are added are segmented polyurethaneureas (PUUs). These polymers are among TPE: The origin of their rubber elasticity is ascribed to their network-like structure formed by the microphase separation between hard and soft segments (see Sect. 3.5.3). The microphase-separated structure of PU and PUU is known to give rise to good blood compatibility which is indispensable for biomedical materials. In other words, the microphase-separated structure is dually important, firstly in displaying rubber elasticity and secondly in showing bioactive properties. About the effect of the morphology of multiphase-structural polymer on the blood compatibility, the concept by Lyman et al. [121] has been widely accepted in the field of biomedical material science: The albumin in blood plasma is supposed to be easily adsorbed on the surface of microphase-separated structure between hydrophilic hard domain and hydrophobic matrix, and the adsorbed albumin is effective in delaying the adhesion of platelets, which lead to the antithrombogenicity.

A group of segmented polyurethaneurea (named SEUU here) was synthesized in our study by using a novel-type prepolymer which was synthesized by introducing hydrophilic polyoxyethylene (POE) segments at the both ends of polyoxytetramethylene (POTM, which is hydrophobic and commercially available). It was assumed that hydrophilic POE units attached at the both ends of POTM may shift the balance of hydrophilicity and hydrophobicity to the direction favorable to biocompatibility. The synthesized SEUU was subjected to the evaluation of mechanical properties and blood compatibility [98, 100, 103, 105]. The introduction of POE segments to the ends of POTM segment resulted in the improvement of antithrombogenicity, i.e., SEUU is better than the conventional segmented PUU in terms of biocompatibility, which was evaluated by in vivo short implantation tests as shown in Fig. 3.17 as well as by in vitro Lee-White clotting tests and in vitro

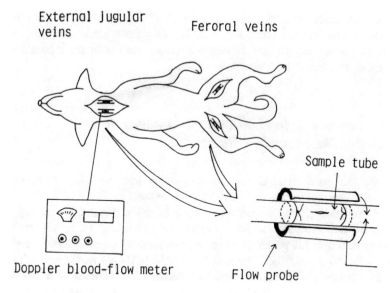

External jugular veins

Feroral veins

Sample tube

Doppler blood-flow meter

Flow probe

Fig. 3.17 Measurement of the patent time of four sample tubes by the short-term implantation in canine peripheral veins. (From Fig. 4 in Ref. [98])

platelet adhesion tests [98]. The wet samples of SEUU which were soaked in normal saline solution at 37°C for two weeks showed lower stresses up to the high strain than segmented PUU without POE segments. That is, the lower moduli mean that the mechanical modulus of the former is found to be similar to that of the living soft tissues compared with that of the latter.

One of the favorable factors of SEUU for improving the biocompatibility is ascribable to the increased hydrophilic nature of the surface of SEUU, which led to the decrease of its interfacial free energy, further resulting in the difficult absorption and easy release of proteins. The hydrophilic and hydrophobic balance of the SEUU surface may have resulted in better blood compatibility and tensile modulus. These results suggest that the more complete microphase separation of SEUU gave rise to the better antithrombogenicity and the lower tensile moduli, which may be explained by the "capping control" mechanism on the surface of multiphase polymers [122].

In Japan, the transplants of organs have been legalized, but the required high cost is prohibitive to new clinical operations. (The cost may include both cost and time for obtaining official approval). As a result, the development of artificial organs has remained to be minimal. The new prepolymer for SEUU mentioned above should have been synthesized much cheaper. However, developmental works on the external artificial hearts, in particular, those for the temporary use, have been actively conducted and recently commercialized. Therefore, a demand for bio-compatible PUs and PUUs is still high as raw materials for "tentative" artificial organs to be used before the operation of organ transplantation.

The medical service using rubber materials having good blood compatibility and adequate mechanical properties similar to the soft tissues would continue, and it is hoped that increasing number of medical researchers are to work on rubber for developing soft biomedical materials.

3.3.3 Elastomers for Lithium-Ion Conduction and the Secondary Battery

In order to meet the global environmental and resource problems, the next-generation sustainable cars have been highlighted since the 1990s. This global trend is suggesting that we are now living in the transportation society, while looking forward to the Information Age [94]. And, presently the most favorable and practical candidate of zero-emission car is an electric car (electric vehicle, EV). One of the authors has described that a fuel-cell vehicle (FCV) would not be ready to meet the demand until the 2020s [94]. Concerning EVs, it does not need to say that the most important issue is the secondary battery, that is, the rechargeable battery of higher performance is essential from the practical point of view. Further, among various batteries, lithium-ion battery is one of the most reasonable candidates, and for its betterment, the development of a good solid electrolyte for the battery has been discussed extensively [96].

Among various solid electrolytes, polymer electrolytes have so far been regarded as very useful materials for the lithium-ion battery, which enable us to produce small, thin, and lightweight devices [96]. At first, a poly(siloxane) system, whose T_g is the lowest among amorphous polymers, becomes a candidate for a molecular design of highly ion-conductive matrix on the basis of the micro-Brownian motion. However, polymers such as poly(siloxane) are nonpolar and it is difficult to dissolve salts in the matrix. Therefore, combinations of poly(siloxane) with polymers in which ion salts are dissolved have been investigated. Accordingly, polymer solid electrolytes with various architectures such as block and graft copolymers have been studied, where poly(siloxane) and poly(oxyethylene) (POE) were combined [106]. However, cross-linking was found necessary for poly(siloxane)/POE copolymers in order to afford sufficient mechanical strength. Unfortunately, cross-linking results in the decrease of ion conductivity, which should be reasonable since diffusion of ions is limited by network structure formation by the cross-linking reaction.

On the other hand, the copolymerization method has generally been utilized to produce an amorphous rubber state in rubber industry. For example, like EPM and EPDM (ethylene–propylene copolymers), epichlorohydrin rubber (ECO) is a random copolymer of ethylene oxide (EO) and epichlorohydrin (EH), P(EO-co-EH) where the ratio of EO/EH is about 1/1. (See **Remark 1** for EPM, EPDM, and ECH.) ECO of high molecular mass of the order of 10^6 shows good mechanical properties without any cross-linking. However, ECO doped with lithium perchlorate showed ion conductivity of only lower than 10^{-5} S/cm at room temperature [107, 111]. Even if the presence of chloromethyl groups was effective to prevent the crystallization of POE to give elasticity to ECO, the group was not considered to contribute to the conductivity increase.

Next, a branched polyether (TEC) in which the oxyethylene segments were introduced by copolymerization technique instead of chloromethyl groups was studied [108–110, 112, 114, 116]. The three oxyethylene segments at the side chain (ETT-3) have effectively stood against the way to crystallization of the POE main chains. Additionally, the POE segments at the main chains and tri(oxymethylene) segments at the side chains are both advantageous to dissolve the salt and to promote the diffusion of lithium cations. When TEC of molar mass of 10^6 was doped by lithium perchlorate at the concentration of [Li]/[-O-] = 0.05 ([-O-] designates the concentration of oxyethylene unit, $-CH_2CH_2O-$), the resulting polyether electrolyte showed one of the highest ionic conductivities as shown in Fig. 3.18 [117]. Film formable inorganic/organic complexes were also prepared from high ionic conductive Li_2S-SiS_2-Li_4SiO_4 oxysulfide glass and TEC [113, 115]. These are examples of the polymer electrolyte design that has incorporated two factors, an

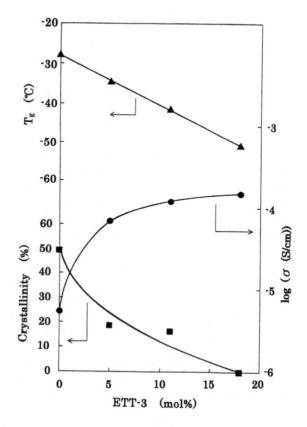

Fig. 3.18 18 Effect of TEC concentration in the copolymer on the crystallinity, ionic conductivity (σ) and T_g of TEC films doped with lithium perchlorate, [Li]/[-O-] = 0.05. (From Fig. 8 in Ref. [117])

amorphous and lithium-ion conduction matrix. However, the polymer electrolyte of lithium-ion conductivity over 10^{-3} S/cm at room temperature is yet to appear.

This uniquely specific field in rubber science has further possibility of being developed into the new arena, solid electrolyte materials for next-generation lithium-ion battery, the capacity of which is large enough to be loaded on EV or PEV (plug-in EV). Possibly different from bioactive arena, polymer solid electrolytes may include another possibility by using soft nanocomposites. Namely, a mixing rubber with nanofiller has been rather traditional for rubber reinforcement, yet it is suggesting unique soft ion-conducting materials [123], one example of which is already mentioned in Sect. 3.3.1 (Fig. 3.16). Such development in rubber field may extend the functional materials more in a near future.

It also can be an important contribution to the low carbon society in accordance with sustainable development of us, which may be additional but essential contribution from rubber side together with that to the improved performance of pneumatic rubber tires.

3.4 Crystallization of Natural Rubber

3.4.1 Molecular Background and Analysis of Crystal Structure

Rubber is not destined to crystalize in general, since rubber molecules are in random motion at room temperature. Or more precisely, at room temperature a rubber molecule is incessantly moving at microscopic level due to micro-Brownian motion of its segments above T_g, even when not in motion at macroscopic level (i.e., standing still). In other words, rubber is fundamentally amorphous and is liquid-like (a polymeric solvent, see Sect. 1.3 and Refs. [95, 96, 117]), though its viscosity is very high. The high viscosity affords apparently solid-like nature to a solid raw rubber, which is actually a viscous liquid. Cold flow displays this hidden nature of rubber visibly: By standing a lump of raw rubber such as SBR, which is not a crystalloid, on a table at room temperature, it deforms and ultimately spreads over the table exactly like spilled water with the lapse of time (in a few months or in a few years, depending on its viscosity).

On the other hand, natural rubber (NR) is uniquely stereo-regular at its molecular level: Only 2 (or 3 depending on the plant species from which NR is obtained) of the monomeric units at a chain end are of *trans*-1,4 configuration, and all the others are of *cis*-1,4 [124]. If we assume a NR molecule whose degree of polymerization (DP) is 1000, the 998 units are of *cis*-1,4 configuration except two *trans*-1,4 units at a chain end. That is, the stereo-regularity of NR is 99.8% (in Sect. 1.2.2, this value was written as 100%). So far, the effort by the synthetic polymer chemists has not attained such a high stereo-regularity in industrial manufacturing of *cis*-1,4-polyisoprene. Additionally, more important point is the presence of the isomeric *trans* units only at one chain end. In other words, the rests of monomeric units are all *cis*-1,4 to result in the sequence length of *cis*-1,4 being 998 which constitutes one *cis*-1,4 sequence [125, 126]. Due to this sequence length, NR shows a uniquestrain-induced crystallization (SIC) behavior of NR, which may be the origin of superior performance of NR over lots of synthetic rubbers [94, 125–128].

Even when a synthetic organic chemist is successful in attaining 99.8% stereo-regularity by chemical polymerization of isoprene in a near future, the two *trans* units are distributed statistically (in accordance with the Markov statistics, or most probably simple Markov, i.e., in random) as usual for many chemical (non-biochemical) reactions [125, 126, 129]. Therefore, the average length of *cis*-1,4 sequences in the synthetic *cis*-1,4-polyisoprene whose DP is 1000 is to be 998/3 = 333, which means the decrease of *cis*-1,4 sequence length to one-third. Since the length of stereo-regular sequence is one of the important molecular bases for polymer crystallization, the drastic decrease of sequence length is less favorable to its crystallization. Due to the high stereo-regularity with a long stereo-regular sequence length, NR is unique in its molecular character as a crystalloid among rubbers which are basically amorphous.

Crystallization was earlier observed on raw NR and NR products by their turning hard and brittle during the winter time in North America and in Europe [94]. This phenomenon is now called low-temperature crystallization of NR, and it had been the main obstacle for the popularization of the NR utilization before the invention of rubber vulcanization in 1839 by C. Goodyear (see Sect. 2.2.1). To tell an essence of vulcanization, it was an unintentionally found method to prevent NR from crystallizing even in winter. The other type of crystallization of NR is mentioned in Sect. 1.3.1 as SIC, which is typically observed in cross-linked NR upon stretching to several hundred percentages of the initial length. The crystal structures of the both are identical, but the two crystallizations observed on NR are much different in terms of their crystallization rate, orientation of the crystallites, and the significance from the viewpoint of materials science. Before going to these features of NR, a brief account on the crystal structure analysis is given here.

One of the most fundamental techniques of determining crystal structure is wide-angle X-ray diffraction (WAXD) method, and its base is the Bragg condition,

$$2\,d\sin\theta = n\,\lambda, \tag{3.30}$$

where d is the distance between the crystal lattices, θ is an angle of incident X-ray to the lattice face, λ is the wavelength of X-ray (for CuKa ray, $\lambda = 1.542$ Å), and n is the ordinal number of the reflection (zero or a positive integer). When the reflections from lattice faces are in the same phase (the Bragg reflection), diffraction is observed in accord with the Bragg condition, which allows us to determine the internal lattice distance d. As already noted in Sect. 1.3.1, WAXD of amorphous materials displays isotropic ring-shaped halo suggesting being amorphous. When not halo but clear ring is observed, crystallization is suggested, but the orientation of the crystallites is random. For X-ray and neutron diffraction and scattering in general, see two recent books, a concise one [130] and an exhaustive one [131].

In understanding the crystallization of polymers, degree of crystallization is one of the key ideas [132, 133]. A polymer is apparently consisting of one long linear chain of high molar mass, and exists in a randomized coil-like conformation not only in solution but also in bulk. Hence, its diffusion coefficient is small, and it is geometrically entangled by the other polymers. These factors are not favorable to crystallization, which of course means the formation of highly ordered structure. In fact, the optimal degree of crystallization among polymers is between 30 and 80%, and the rest remains amorphous. In the case of polyethylene (PE), single crystal (an almost perfect crystal, i.e., nearly 100% crystallinity) may be prepared, which is due to its simple chemical structure $-(CH_2CH_2)-$ and to the fact that its range of possible crystallization temperature is wide, i.e., $(T_m - T_g)$, is approximately wider than 200 °C. Thus, crystallites are generally coexisting with surrounding amorphous regions in polymers. Based on this fact, a structural study on the coexisting ordered and disordered states is an important subject in polymer science, which is often called polymer morphology. Determination of degree of crystallization (the first step of the morphological study of polymers) has been done by several methods. Among them four popular ones are from specific volume, X-ray diffraction, differential scanning calorimetric, and infrared absorption measurements [131, 133, 134]. It is highly advisable to use at least two methods for determining the degree of crystallization, preferably including WAXD measurement as one of them.

Generally speaking, some specific conditions are mandatory for a polymer to attain a high degree of crystallization: In order to obtain good fiber materials, high-speed (uniaxial) spinning followed by drawing is a standard. In the preparation of PE single crystal, a dilute solution of PE is kept under a very mild stirring for long. These examples may suggest that amorphous might be a normal state of polymers since the macromolecularity (being of high molar mass, see Sect. 2.3.3 for this word) is not a favorable condition for crystallization, even though amorphous may thermodynamically be a metastable state.

It is well known that X-ray study on NR has pioneered the structural research of polymers, which is partially described already in Sect. 2.3.4. One of the most popular textbooks of polymer science has used NR data to explain polymer crystallization behaviors. Contrary to PE, however, the range of possible crystallization temperature, $(T_m - T_g)$, is 90 °C, much narrower for high degree of crystallization. This introductory subsection is followed by two subsections, one describing SIC, unique crystallization behavior of the cross-linked NR, and the other is low-temperature crystallization of NR. Note that the latter is not usually observed in tropical countries where NR is tapped and collected due to hot climate.

3.4.2   Strain-Induced Crystallization: Template Crystallization

(1) Time-Resolved Measurement

Unique crystallization upon stretching, i.e., strain-induced crystallization (SIC), of NR was reported by Katz as early as in 1925 [135], which is already noted in Sect. 2.4.3. He got so-called fiber-like X-ray patterns, and he described SIC as "fibering." This could be a trailblazing study in polymer crystallization: He seemed to have regarded NR as a macromolecular entity, and more consideration of the results might have suggested him crystallization onto the fully extended NR chains. Unfortunately, however, he seemed to be too much talented to further continue the pioneering SIC study on NR, and shifted his research interest to various areas [136]. Due to the very high rate of crystallization in SIC, scientific treatments of SIC have been very difficult until the advent of a powerful synchrotron radiation as the X-ray source, which enables the time-resolved measurement of SIC.

By using the synchrotron radiation as the X-ray source, recent time-resolved studies on SIC behaviors of cross-linked NR suggest that the characteristic high stereo-regularity of NR falls into full advantage in its elucidation of SIC behaviors: By uniaxial elongation of the cross-linked NR sample to a few to several times of its original size, some relatively short network chains are fully extended, and the NR sample is under a dynamic condition to result in spontaneous and instantaneous crystallization onto the extended *cis*-1,4-polyisoprene sequence [125, 126, 137–145]. Cross-linked NR is surely an amorphous material, yet its unaffected ability of SIC gives rise to self-reinforcement by the resulted crystallites which actually are in situ generated reinforcing nanofillers in the rubbery matrix. Here, the macroscopic deformation of the sample is assumed to be microscopically reproduced in the deformation of rubber network chains, i.e., the affine deformation is assumed.

In Table 3.5 are shown the compounding recipes of NR and IR (for IR, synthetic analog of NR, see Sect. 1.2.2) for vulcanization. IR itself does not contain non-rubber natural components such as proteins, phospholipids, and fatty acids. It is of value as a control sample of NR for comparison. The cross-linked samples were subjected to SIC measurements at BL40 XU beamline (X-ray wavelength is 0.1 nm) of SPring-8, Hyogo, Japan, or at X27C beamline (X-ray wavelength is 0.1366 nm) of NSLS, Brookhaven National Laboratory, New York. Handmade uniaxial tensile tester was set up at the beamline, and simultaneous WAXD and tensile measurements were carried out on the sample under elongation. Use of the synchrotron radiation enabled us to take WAXD images of the sample in every several seconds into a computer, and time-resolved measurements of structural and mechanical data have been established [137–142]. References [143–145] are review papers on SIC of NR.

In Fig. 3.19 is shown the progress of SIC upon stretching of NR and IR vulcanizates [143]. Changes of CI (crystallization index, a measure of degree of crystallization) and OAI (oriented amorphous index, a measure of network-chain

Table 3.5 Recipes for vulcanized NR and IR samples in phr[a]

Sample code	NR-1	NR-2	NR-3	NR-4	NR-5	IR-2	IR-3	IR-4	IR-5
Rubber[b]	100	100	100	100	100	100	100	100	100
Stearic acid	2	2	2	2	2	2	2	2	2
ZnO	1	1	1	1	1	1	1	1	1
CBS[c]	3	2	1.5	1	0.75	2	1.5	1	0.75
Sulfur	4.5	3	2.25	1.5	1.125	3	2.25	1.5	1.125
Time[d] (min)	10	12	12	14	18	18	21	25	30
$v^e \times 10^4$	2.12	1.78	1.46	1.31	1.01	1.66	1.36	1.29	1.03

From Table 1 in Ref. [143]
[a]Part per one hundred rubber by weight (except time and v^e)
[b]For NR and IR, RSS No. 1 and IR2200 were used, respectively
[c]N-Cyclohexyl-2-benzothiazole sulfenamide
[d]Time for vulcanization at 140 °C
[e]Network-chain density in mol/cm^3

orientation) are plotted against stretching ratio: The sample code is that shown in Table 3.5, and the larger is the code number, the smaller is the cross-link density, i.e., network-chain length is larger [143]. Results of (**c**) and (**d**) show that orientation of the network chains starts immediately upon stretching and it increases by a constant slope in IR while in NR the slope shows dependence on network-chain density. Here, stretching ratio α is defined as $\alpha = l/l_0$, where l_0 is a sample length before deformation and l is that under stretching. In NR, SIC was observed in the vicinity of the ratio of 3.5, depending on the network density (see (**a**)), while SIC of IR started at the ratio of 4.5 independent of the density (see (**b**)). Deformation was discontinued before their rupture, and let them contract at the same speed to the original length. In both samples, the crystallites have melted with the decrease of strain to recover the complete amorphous state at the original point [139, 140].

Even in IR, the stereo-regularity of which is lower than NR, SIC was observed [140]. This finding may suggest that the green strength (tensile strength of rubber compounds before vulcanization) has not much to do with SIC, since the green of IR is much inferior to that of NR. However, SIC behaviors of raw or not cross-linked NR or IR are much more complicated than those of cross-linked ones, and the exact conclusion is still to be elucidated. Some of the differences between NR and IR may also due to the difference in heterogeneity of network structure as well as the presence or absence of non-rubber components. Onset strain, where crystallization starts, in NR around 3.5 depending on the network-chain density and in IR at 4.5, is explained by considering melting point increase per the consideration by Yamamoto and White [143, 146].

The effect of stearic acid on SIC is also investigated [141]. NR contains a small amount of stearic acid as an original natural component, and it has been known that stearic acid is functioning as a nucleating reagent in low-temperature crystallization. However, IR vulcanizate without any stearic acid did show SIC behavior similar to that shown in Fig. 3.22. Accordingly, SIC is not initiated by stearic acid which is

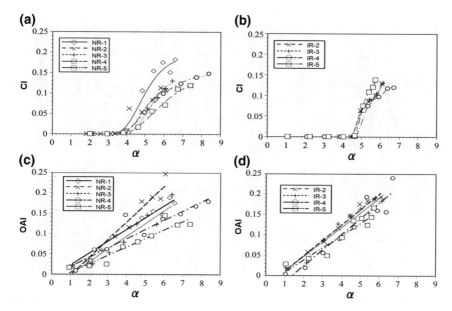

Fig. 3.19 a, b Advances of SIC and **c, d** orientation of amorphous part by uniaxial elongation of **a, c** NR and **b, d** IR vulcanizates (from Fig. 5 in Ref. [143])

active as a nucleating agent for low-temperature crystallization as explained later in Sect. 3.4.3.

(2) **Template Crystallization: Mechanism of SIC of Cross-Linked Natural Rubber**

Taking the findings above described into account, a possible mechanism of SIC of NR has been proposed [139–141, 143, 144], as shown in Fig. 3.20. Upon stretching, some network chains of a certain length (possibly a short one, but at least the length of several to ten monomeric units) are fully extended at a high strain, and they serve as a template of crystallization of the nearest amorphous network chains. In other words, crystallization onto the template seems to start spontaneously, leading to the crystallites which are highly oriented to the stretching direction. The threshold strain value of onset of SIC was found around $\alpha = 3.5$–4.0 for NR and $\alpha = 4.5$ for IR.

It is notable that a certain network chain is fully extended upon stretching at a certain elongation, i.e., formation of the template is a matter of inevitableness when a cross-linked NR is subjected to higher stretching. And, the crystal growth on the template would possibly be depending only on availability of the longer network chains around it. In this case, even one single fully extended network chain of a certain length might be considered to be a template, and it is different from the so-called nucleus whose formation has to be statistically treated, and its formation is often described as "sporadic." If a few or several fully extended network chains

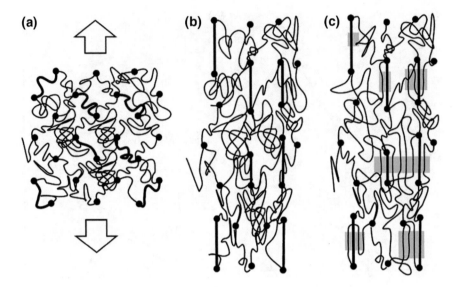

Fig. 3.20 Crystallization schemata on SIC of cross-linked rubber. Relatively short extended chains are drawn as thick lines. Filled circles represent cross-links. **a** Before deformation. **b** After deformation just before the onset of crystallization. **c** The fully stretched chains have functioned as a template of crystallization. Shaded parts show crystallites formed under elongation (from Fig. 4 in Ref. [143])

were present side by side to form a bundle (equivalent to shish part of the shish-kebab crystals) as assumed in a report [147], it might have possibly been considered to be a nucleus. However, such a bundle formation of fully extended network chains of a certain length is virtually impossible due to the improbable occurrence of the fully extended network chains at side by side, together with the highly limited mobility of the extended network chains in the network structure. Also, it has to be noted that the network-chain length, i.e., molar mass distribution of network chains produced by vulcanization, is known quite wide, definitely not at all mono-dispersed. At a certain elongation before the onset strain α, the assumed bundle formation of the fully extended network chains (assumed in Ref. [147]) is absolutely improbable. It may be also notable that the length of the network chain is to be larger than a certain value in order to function as the template of crystallization. Though the exact length for the template to initiate the crystallization is not known yet, the length of dimer or trimer level would not effective for crystallization, and the length of several to a few tens monomeric units may be the minimum for the template.

In addition, SIC may uniquely be different from crystallization under orientation in general, in terms of the nature of the strain, i.e., uniaxial stretching or elongation. Orientation of polymeric chains is usually conducted by applying shear [148] using compressional strain (like in calendar, compression, extruding, or injection moldings), not by stretching using tensile strain. Interestingly, the stretching mode may naturally be observed in the spinning of silk by a silkworm and/or the making of a silk fiber by a spider. In the former, rapid movement of the worm's head is for the spinning by stretching the silk fiber in order to prepare a cocoon and in the latter, motion of the spider is in a fraction of a second [149] to construct a network chain. Namely, their methods of spinning are not by shear: Their spinning does not seem by applying the pressure from their internal organ toward their mouth to extrude a fiber, but by stretching of the fiber outside of their body. Resemblance seems to be worthy of further comparative studies of SIC of NR with them.

The lattice constants a, b and c that define the unit cell size, calculated assuming Nyburg's crystal lattice [150], have shown that the constants a and b are decreased while c (along the stretching direction) is increased upon stretching. It is suggested that the macroscopic stress is transmitted to the microscopic lattice cell. The behavior of a, b and c is qualitatively similar to that of polyethylene (orthorhombic, with the c-axis along the direction of the planar zig-zag chain backbone) by temperature increase, which may be due to the anisotropic change by the thermal expansion since c-axis is the direction of planar zig-zag chain backbone, and the deformation of lattice may rise from the surface strain effect [151, 152]. The product of a, b, and c gives the volume of the lattice, which is found to decrease with elongation on cross-linked NR samples. Rubber has a Poisson ratio of 0.5, which means macroscopically no volume change on deformation, i.e., being incompressible. The volume decrease of NR lattice is unique, and this may support a pantograph deformation model of network structure [139, 143, 153–155] as shown in Fig. 3.21 [143]. The schemata are assuming numerous units consisting of several network chains in the network structure, and the macroscopic deformation is due to their relative geometry but not their sizes. This model **c** may qualitatively not be incompatible with the recent study on dynamics [156], and worth being considered much more for describing deformation behavior of a heterogeneous network structure, the typical of which should be rubber vulcanizates [157–159]. As described in Sect. 4.1, structuring of nanofillers like carbon black results in network-like structure formation in rubber matrix (e.g., see Fig. 4.4). Model **c** may be the case for the nanofiller-loaded rubber as well as for the networks of nano-fillers. This hypothesis is still to be proved, and further studies are now in progress.

Summing up the mechanism elucidated so far, SIC of NR, is specific kind of crystallization onto a template, and the crystallization is spontaneously recognized once the template (an extended network chain of the length) is formed upon uniaxial elongation of the sample. This template formation is not sporadic, but it is a matter of necessity by uniaxial tensile elongation. This crystallization mechanism, template crystallization, may be qualitatively not contradictory with a recent report on segmental dynamics [156]. In polymer crystallization by nucleation mechanism (homogeneous crystallization), the start of crystallization is dependent on statistical fluctuation dynamics, and the presence of a certain size of nucleus is required, which is the case even under shear flow in the crystallization of isotactic polystyrene [160]. The beginning step of SIC of NR seems to give rise to a unique new arena of polymer crystallization: The first step is the appearance of a fully extended network chain of a certain length, which functions as a template to induce

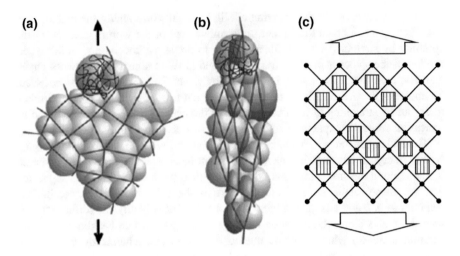

Fig. 3.21 a and **b** Schematic representation of deforming cross-linked rubber, and **c** the pantograph model to explain the deformation including crystallites generated in situ by SIC in the rubbery matrix . *Spheres* designate globules of coiled rubber network chains, where *lines* indicate the end-to-end direction of the network chains. Before elongation, all globules are spherical (**a**), but even after the deformation, only a part of the globules are elongated which are drawn in dark (**b**). For explaining the tensile behaviors of cross-linked rubbers, the pantograph model may be attractive (**c**). Here, the crystallites by SIC are located in the frames of surrounding chains, the deformation of which may transmit the compressive stress to the crystallites (from Fig. 7 in Ref. [143])

crystallization of nearby amorphous network chains, followed by in situ crystal growth to a crystallite. It is notable that a fully extended network chain appears by stretching the cross-linked NR to a certain elongation, which is a logical result of the relationship of cause and effect (the causality rule). On the contrary, the nucleation is due to fluctuation, and its formation is sporadic or by chance to be explained by the stochastic process. The big difference of the crystallization rate between SIC (of less than 1 s order) and low-temperature crystallization of NR (of more than years order; see next Section, Fig. 3.24) suggests the completely different mechanism of the two crystallization modes of NR.

This crystallization, SIC of the cross-linked NR, may therefore be better called "template crystallization", accommodating the possibility of a single molecular template [161]. The template formation is not at all stochastic, while nucleation in low-temperature crystallization of NR and in almost all crystallizations, except heterogeneous crystallization in which a nucleating agent is annexed, is a fully stochastic process. It is interesting to note that the proposed template crystallization here is in essence very similar to so-called template polymerization in which a mother polymer is fixed as the template on the solid surface and on which monomers are adsorbed, followed by polymerization to produce the next-generation polymer. The son or daughter polymer has often exactly of the same structure or

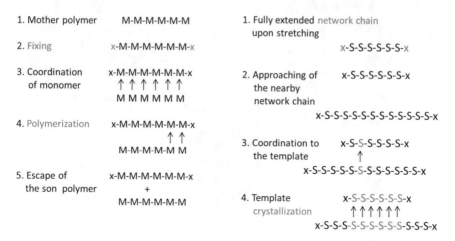

Fig. 3.22 Comparison of template crystallization with template polymerization. M, monomer; S, the Kuhn segment; x, fixing or cross-linking point; *Arrow* shows some attractive interactions (hydrogen bonding, or van der Waals or London's dispersion force)

properties of the mother polymer. This polymerization technique has its origin in the invention by R.B. Merrifield in the polypeptide field, who won the Nobel Prize in Chemistry for his solid-phase peptide synthesis in 1984 [162, 163]. Also, this is possibly the normal or commonplace in vivo technique observed in the synthesis of DNA (deoxyribonucleic acid) of the next generation to produce the complementary DNA of the parent. The SIC process is therefore represented as shown in Fig. 3.22 in comparison with that of template polymerization. The template crystallization in Fig. 3.22 can be the first starting model of the further analysis of SIC. For example, kinetic study based on this model is expected to elucidate the SIC mechanism further in details. There, the step 2 and/or step 3 may be the rate-determining step. In other words, the Brownian motion of the nearby network chains approaching the template may be the slowest hence rate-determining step [161]. Some much sophisticated technique equivalent to pressure jump (P-jump) method is to be planned for the kinetic approach to template crystallization.

SIC, in particular, of NR has recently been actively investigated, especially on arresting fracture of rubber vulcanizates by SIC [164–167]. In other words, SIC is assumed to be effective in improving the fracture toughness of rubber products. Also, effect of cross-link density on SIC [139, 168], SIC behaviors of peroxide cross-linked NR [141, 169], effect of SIC on nanofiller-filled NR [170–173], SIC of synthetic rubbers [174, 175] are reported, as well as SIC of NR vulcanizates from latex [176]. SIC behavior of NR is now focused to be the most important and interesting feature of NR. In 2014, a young physical chemist's Ph.D. thesis was published as a monograph on NR [177], which contains somewhat exhaustive list of papers relevant to SIC.

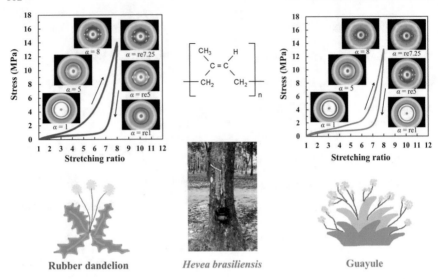

Fig. 3.23 Strain-induced crystallization behaviors of natural rubbers from guayule (*Parthenium argentatum*) and rubber dandelion (*Taraxacum kok-saghyz*). The stretching ratio with "re" means the ratio in return to the original length (Modified the graphical abstract of Ref. [178])

(3) **SIC as an Index of the Alternatives of *Hevea* Rubber**

Most recently, NRs from guayule (*Parthenium argentatum*) and rubber dandelion (*Taraxacum kok-saghyz*, also known as Russian dandelion, Kazak dandelion, Buckeye Gold, TK or TKS) are subjected to the evaluation of SIC behaviors [178] as shown in Fig. 3.23. NR from the two plants has showed more or less comparable SIC behaviors with that from the *Hevea* plant, and are now considered to be good candidates as an alternative of NR from *Hevea brasiliensis* which has been the most valuable source of NR among many rubber producing plants [94, 125–127, 178].

However, the *Hevea* tree as a sole source of NR is supposed to have two potential problems: One is biodiversity [94, 125–128, 179] and the other, biosecurity [94, 125–127, 179–181]. The former is evident: The source of NR has been too much dependent on *Hevea brasiliensis*, which has to be rich in biodiversity in terms of stabilizing the resource. The latter issue is due to localization of *Hevea* trees in tropical Southeast Asian countries (hence, a geopolitical factor [125, 181]) and it also causes pathological problems [125, 126, 179, 180]. Guayule and rubber dandelion are now considered to be promising alternative sources of NR for delocalizing NR resources instead of the *Hevea* [94, 126–129, 179]. The WAXD results of the two NRs under tensile stretching and contraction shown in Fig. 3.23 have suggested them to display the almost comparable SIC behaviors to *Hevea* NR, and to qualify as a good substitute, in terms of their expected physical properties [178]. Further investigation on their properties is mandatory, still displaying SIC behavior is one of the indispensable conditions for the expected higher performance of NR. In other words, the alternative NRs of *Hevea* rubber have better be checked their SIC behavior before the elucidation of their various properties at present.

3.4.3   Low-Temperature Crystallization

SIC is the most characteristic and important feature of natural rubber (NR), or more simply, "NR is standing alone due to the SIC behaviors." Crystallization of NR should have been meaning SIC, due to its utmost importance to NR. Unfortunately, however, "crystallization of NR" has been used to mean "low-temperature crystallizationof NR" until quite recent time. Following the naming of SIC, low-temperature crystallization (LTC) should have been named "temperature-induced crystallization (TIC)," but historically TIC was noted much earlier than SIC, and has been called simply crystallization of NR or low-temperature crystallization of NR. The latter is the title used here.

In an edited book on NR published in 1963 [182], Chap. 9 is entitled "Crystallization in natural rubber" (authored by E.H. Andrews and A.N. Gent), but SIC was not described there. The new version of the book published in 1988 [183] contains Chap. 18 entitled "Low-temperature crystallization of natural rubber," but any chapter on SIC is not included. From the viewpoint of utmost importance of SIC in understanding mechanical performance of NR, it is surprising that a chapter on SIC in NR books is firstly found in the one published as late as in 2014 [145]. The basic reason of this delay is due to the extreme high speed of SIC of NR, while that of low-temperature crystallization rate is slow enough to be studied by conventional techniques: It needs at least days, usually weeks, months, or even years for its clear recognition even in winter in the temperate and subarctic zones. Note again that usually the low-temperature crystallization is not observed in tropical countries. For the scientific analysis of SIC of NR in particular, observation of X-ray diffraction at a modern synchrotron radiation facility is absolutely necessary, e.g., at SPring-8 in Hyogo, Japan. It has been found mandatory to conduct the time-resolved simultaneous measurements of tensile and WAXD, in order to elucidate the details of SIC [137, 139–144, 169, 171, 172, 176, 178], since the rate of SIC of NR is so high to necessitate WAXD measurement of millisecond timescale. This is one of the essential differences from low-temperature crystallization of NR, and the exact kinetic study of SIC of NR should be enabled in a near future for which a new kinetic technique applicable to high-rate crystallization processes is to be figured out.

Taking these situations on NR crystallization into account, the readers who want to study low-temperature crystallization of NR had better read a paper published in 2013 [184], in which both crystallization results are presented and compared. In this paper, the word, temperature-induced crystallization (TIC), is used instead of low-temperature crystallization. The NR samples for TIC study were kept at -10 °C for one month, and the ring-shaped WAXD patterns of all the samples are reported, which suggest random orientation of crystallites which were produced by TIC.

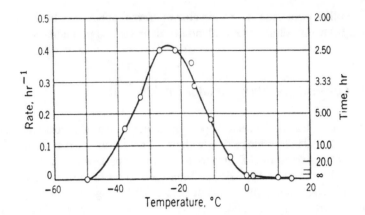

Fig. 3.24 Rate of low-temperature crystallization of natural rubber as a function of temperature (from Fig. 2 in Ref. [186])

Low-temperature crystallization of raw NR may proceed during its storage for a long time, and the NR is better to be warmed for melting crystallites before dumping into a mixer for compounding with curing reagents, fillers, and any other additives. Without preheating, raw NR material may show an abnormal behavior at the initial step of processing [185]. The equilibrium T_m of NR is around 35 °C, which is not much different from room temperature. Due to the small supercooling, it does not seem easy to crystallize. However, T_g of NR is very low at around −56 ° C, and the maximum crystallization rate was observed at −25 °C [186, 187]. Fig. 3.24 shows the well-known results [186], which has been used in many textbooks in polymer science (see Ref. [134] as an example). This is the origin of the expression, "low-temperature" crystallization. For nucleation, "The larger the supercooling, the better for crystallization," that is, crystallization at as low temperature as possible from T_m is favorable. On the other hand, at a lower temperature near T_g, the micro-Brownian motion, hence diffusion, of rubber chains is slow to result in much lower degree of crystallization. Competition of the two factors results in a maximum in crystallization rate versus temperature plot. This rule for temperature dependence of polymer crystallization was first reported on NR [186, 187] and followed by lots of results on the other polymers [130, 131, 133]. Toward rubber scientists, chloroprene rubber (CR) is particularly notable among them: Its maximum temperature of low-temperature crystallization is −10 °C [188].

Without the addition of any nucleating agent, homogeneous nucleation is generally a precondition of crystallization for polymers, which is depending on statistical fluctuation at molecular level [129, 131, 186, 189, 190]. While the crystal growth rate is dependent only on temperature, nucleation rate is strongly dependent on various experimental conditions such as previous temperature of melting, cooling process from the molten state, heating process from the glassy state, etc.

[190]. The nucleation is therefore expressed as being sporadic, and further mechanistic studies on the puzzle solving are recently in active research. For example, molecular dynamics simulation is extended up to 100 ns to investigate the evolution of cluster size, intrachain vs. interchain potential energies and pair correlations of PE solution at 300 K: The embryonic development begins with the aggregation of *trans*-rich sequences of characteristic length of 2 nm, forming clusters of short stems which reorganize to the embryonic clusters, and these clusters may give rise to nucleation [191]. Consequently, scientific considerations are to include statistical treatments [126, 130]. The fluctuation may exactly be treated in rheology as a viscoelastic behavior of a polymeric chain using the reputation theory. No such attempts for explaining the nucleation process have been reported as far as we know. When a nucleating agent is added to the crystallizing system, it is named inhomogeneous or heterogeneous crystallization, which has been a popular technique to reduce the crystal size by producing as many nucleates hence crystallites as possible.

The detailed crystal structure of NR, which is the final crystal both via SIC and via TIC, has been studied by Mayer and Mark (see Sect. 2.4.3), Bunn [192], Nyburg [150], Natta et al. [193], Takahashi et al. [194], Immirizi et al. [195], and Rajkuma et al. [196]. However, the final conclusion on a few details is still to be investigated.

On low-temperature crystallization of NR, time to reach half the degree of crystallization to the equilibrium one ($t_{1/2}$) is from several hours to a few days at -25 °C, at which $t_{1/2}$ shows its maximum value. At room temperature, it is between a few weeks and several months or even years. Wood and Bekkedahl [186] wrote at 14 °C the time required for crystallization was of the order of years (see Fig. 3.24). It was reported that the removal of non-rubber components by solvent extraction delays the crystallization, and the addition of long-chain aliphatic acids, e.g., stearic acid, has been known to accelerate it [197, 198]. These results indicate that the low-temperature crystallization (or TIC) of NR is usual homogeneous or heterogeneous nucleating crystallization, different from SIC of NR. Additionally, the maximal degree of crystallinity attained so far was 40% [199], not much higher than that by SIC. Relatively low degree of crystallization is a feature of polymers in general, but in the case of NR, higher chain mobility due to the micro-Brownian motion around room temperature, which is leading to rubber elasticity, is a reason in spite of its very high stereo-regularity.

In the cross-linked NR, i.e., in many rubber products, recognition of low-temperature crystallization may need a few years, or most probably no crystallization at all even after 10 years. Cross-linking is the only or at least the most efficient way to arrest the low-temperature crystallization of polymers in general, as well as NR. In the case of NR, the equilibrium T_m (35 °C) is decreased 20% by introducing cross-linking at 1% of monomeric units [200], which suggests a practical safety of the cross-linked products of NR. Even in cross-linked NRs, however, the low-temperature crystallization has been a trouble-maker in the countries at temperate or subarctic zones: Some NR products, e.g., NR bearings for earthquake isolation (see **Remark 4**), NR blocks, or sheets for vibration control to be placed under bridges and buildings, etc., are standing for over a few decade supporting the heavy weights. They may lose their elastic property by low-temperature crystallization, which gives rise to the damage of their function [182, 183]. Checking the elastic property of rubber products when used under a quasi-static condition is to be practiced regularly.

Some means to prohibit the low-temperature crystallization of NR have been sought for many years, but in vain. Mixing of nanofillers like carbon black and particulate silica is a standardized practice for rubber reinforcement. A paper claimed nanofillers might be acting as a nucleating agent [201], but it was found not to be the case [202]. Nanofillers did not influence to SIC, too [167]. The most effective is the chemical modification of molecular level, which is exactly cross-linking or vulcanization as C. Goodyear clearly showed in 1839 by inventing the vulcanization, though he did not understand the chemical nature of it at all. For example, low-density polyethylene used for insulating electrical wires is effectively cross-linked, e.g., by the continuous process using electron-beam irradiation. These situations are more or less related to the fact that it is difficult or impossible to bring nucleation under control from a theoretical viewpoint. In industry, controlling the degree of crystallization of plastics is normally practiced by utilization of various nucleating agents, which are producing as many crystallites as possible to result in reduction of the size of crystals.

SIC is the most prominent performance of NR products, though it is a kind of anomaly: NR is basically an amorphous material. SIC is one of the keys for understanding high performance of NR, which is used under the most unsparing dynamic conditions for organic materials, e.g., top tread rubber of the aircraft and heavy-duty truck tires. One of the remaining problems on SIC is the lack of fully quantitative kinetic study to elucidate the crystallization processes under stretching. The rate may possibly be too fast to be followed by the tensile tester to be equipped on the beamline at present [203]. To overcome this experimental difficulty, a special kinetic method, for example, a jump method, has to be figured out in order to reasonably quantify the rate of SIC. Instrumentally, is it possible to realize a strain-jump, like T- or P-jump? Here, T = temperature or P = pressure, respectively.

The tendency for low-temperature crystallization of NR is intrinsic due to its high stereo-regular molecular structure and low T_g. However, it is simply trouble making during its storage and processing, and more importantly it proceeds when the vulcanizates of NR are in use as structural and functional materials. That is to say, from practical point of view, low-temperature crystallization is a phenomenon to be avoided as much as possible [126, 182, 183, 188, 199, 204]. It is strange and somewhat regrettable that this simple fact does not seem to have been understood well enough, by some rubber scientists and rubber engineers. To appropriately differentiate SIC from low-temperature crystallization is surely the first step for understanding and efficiently utilizing NR for various purposes from a viewpoint of sustainable development (SD).

3.5   Reactive Processing of Rubber and Thermoplastic Elastomers

3.5.1   Chemical Cross-Linking and Physical Cross-Linking

Generally, polymers in bulk (solid states) possess a random network-like structure with entanglement (due to the topological interaction) of the Gaussian chains, even though the network-like structure formation due to entanglement is temporary and dynamic in nature. Therefore, characteristics of one macromolecule, i.e., molecular properties, are possibly evaluated only at a dilute solution state where isolated one macromolecule is observable, which is impossible in bulk. Of course, this is the case for a rubber molecule, too. When rubber is chemically cross-linked, the entanglements are trapped in network chains. Upon elongation, some of the trapped entanglements are expected to function as a movable yet permanent cross-linking point. Contribution of the trapped entanglement to rubber elasticity is still under active debates. Roughly speaking, the affine network and phantom network theories are the two developments from the classical rubber theory [205], and pro or con to the role of trapped entanglements may be the watershed between the two. Though this debate is theoretically important and urgent, it is not highlighted in the present book. Relevant original papers by us [32–34, 206–208] and advanced textbooks are to be studied [47, 205, 209]. (In this connection, "entanglement" is also a technical term describing a quantum correlation, which J. Bell used in persuading the doubt on or disbelief in the validity of quantum mechanics, disclosed actively by A. Einstein and E. Schödinger. However, it is not much confusing, since the two fields are remote enough not to be confused.)

On the other hand, the three-dimensional network formation by a stable chemical cross-linking such as vulcanization is already described in Sect. 2.2. Between these two kinds of network structures, i.e., the chemically bonded stable one and the temporary one due to physical entanglement, there is one more category of physical networks, which are more stable than those by entanglement but less than those by covalent bonding. In this section, such a network, thermoplastic elastomer (TPE) [210], is introduced. The network structure of TPE is due to the physical association, not due to the chemical bonding. Characteristics of the three kinds of polymeric network structure are summarized in Table 3.6 [95, 211].

In a TPE molecular chain, two segments are recognized: a hard segment and a soft segment. The hard segments aggregate together by some physical interactions, e.g., Van der Waals force, hydrogen bonding, ionic interaction, to result in the formation of hard-segment domains in the soft segments as a matrix. The hard-segment domain works as a physical cross-linking site in the soft-segment matrix, to form a network structure. From a viewpoint of time, the topological entanglement is temporary and changeable, whereas the hard-segment domains are stable. In terms of temperature, however, the hard-segment domains are melted at a high temperature, followed by softening of TPE to result in the disappearance of the network structure. Namely, TPE is thermoplastic (similar to many plastic materials

Table 3.6 Three kinds of network structure of polymer

Type	Cross-link	Structure	Examples
Covalently bonded network	Permanent, Localized	Point	Rubber vulcanizate, End-linked polymer network, Thermoset resin
Reversible network	Temporary, Localized	Point, Domain, Molecular bundle	Thermoplastic elastomer, Biopolymer gel
Entangled network	Temporary, Delocalized	Topological constraint	Raw rubber, polymer melt, Concentrated polymer solution

From Table 1 in Ref. [210]

except thermosetting resins), which is an origin of its name "thermoplastic elastomer". In Table 3.6, the two network structures, except chemically cross-linked ones including rubber vulcanizate, are due to a physical origin, namely, due to physical cross-linking. A chemically cross-linked polymer has often been named as a gel, and nowadays, a chemically cross-linked one is a chemical gel, and a physically cross-linked one is a physical gel. Note that the gel means the solvent-swollen networks in general. Since the cross-linked rubbers and TPEs described here are usually used under a dry state, not under a solvent-swollen state. Therefore, the word gel is not used in this book.

3.5.2 Rubber Processing from a Chemical Standpoint

(1) Design of Rubber Processing

Rubber processing generally means a technical process to make a product from raw materials, where some specific engineering techniques are requested. In the rubber processing, a good design is absolutely important for the relevant industries, since the rubber products are necessarily a cross-linked composite of several or often more than ten components. Here, therefore, the rubber processing is described from a viewpoint of design [95, 212]. At first, a product design should be well planned at the beginning. Namely, the processing cannot be started without any product design, however, rough it may be. It is also noted that the product design may be subjected to modification or change according to the results of each processing step, if necessary. Bees can build up their nests with an excellent honeycomb structure instinctively. However, human beings have to start the processing from the product design even if it may a tentative one, because we do not have inherited such an instinct as the bees have. Some talented engineers might not need to write down a design. In that case, however, they have had something inside their brains that enables them to work logically and rationally according to his or her design. They are not bees, and they are conscious of the design and do have a certain design somewhere in mind.

(2) Compounding Design

In practice, the first step of rubber processing is a compounding, where a selection of various raw materials and their amounts should be planned. For the selection of raw rubber, a molecular design (i.e., design of a chemical structure of rubber) might be necessary, even if you prefer a commercially available rubber: Commercial ones are usually offered in several grades in accordance with the customers need. In the case of rubber blends, which are often utilized in rubber industry, product designs not only for the chemical structure of rubbers but also of the higher-order structure of blended rubbers are required. Here, the rubber blend means a preparation of raw rubber material by mixing two or more different rubbers. Next, necessary reagents and their amounts are planned depending on functions and performances requested for the final rubber products.

In the rubber processing, design of vulcanization is the most important. That is, a recipe for curing reagents is to be determined: Sulfur, organic accelerators, activators, a prevulcanization inhibitor (PVI), and processing aids, etc., are carefully selected. In practice, preliminary experiments are usually conducted for determining a suitable vulcanization system and conditions. Excellent compounding recipes are often treated as confidential and are not disclosed at all. Smart and highly workable sense on the basis of past experiences still governs the compounding design. Rubber technologists have to pay their efforts to scientifically analyze their senses as possible as they can, in order to systematically classify their recipes for the future utilizations. For this aim, careful study of one or two technical books is highly recommendable [181, 182, 188, 213–216].

(3) Mixing and Mechanochemical Reactions

Once the compounding recipe is fixed, how to mix and which mixer is to be used under what kinds of conditions have to be determined. That is, the next step is to design how to mix raw rubbers with compounding ingredients to get compounds for vulcanization. Before mixing the ingredients with rubbers, a mastication of raw rubbers is necessary for softening the raw rubbers in order to make the mixing easier, which is achieved by mechanical milling under shear stress. For NR, this process is absolutely necessary for good processing, and some unusual behaviors of NR are reported [185]. The mastication and also mixing rubber with any ingredients may probably involve the cleavage of rubber chains to bring about the decrease of molar mass of rubber. Namely, the chemical reaction can possibly occur by the mechanical shear, which is named mechanochemical reaction. (Mechanochemistry is a branch of chemistry on the chemical reactions induced mechanically, or that on transformation of mechanical energy to chemical one. Conversely, chemomechanics is on transformation of chemical energy to mechanical one.) The mechanochemical reaction is particularly noted in cold mastication of NR [217–219]. During mixing, it is often observed that rubber chains are subjected to cleavage and the resultant fragments react with each other and with the mixed ingredients or with a reactive molecule intentionally added. In the former case, the mechanochemical reactions observed before the vulcanization step are called "scorch," and the scorch has to be avoided. In the latter case, when vinyl monomer, such as methyl methacrylate (MMA) is purposely premixed, PMMA might be grafted onto rubber, to result in the chemical modification of rubber. NR grafted with PMMA may be a kind of thermoplastic elastomer (TPE) described in the next subsection.

Also, during the mixing step, an addition reaction and/or a chemisorption of rubber chains occurs on the surface of nanofillers such as carbon black (CB) followed by a formation of bound rubber on the surface of the nanofiller. The formation of bound rubber by the mechanochemical reaction has been considered to be one of the factors for reinforcing effect of rubber by reinforcing fillers (so far, CB and silica are typical examples). Since the bound rubber is difficult to be removed from the fillers even by extraction with a good solvent, it is also named as a filler gel. Especially, in the case of CB, rubber engineers often use the word "carbon gel" for the bound rubber on the surface of CB. However, the term of carbon gel is not used in this book, because it is confusing with a network-like structure of CB particles which may be formed in the rubber matrix (see Sect. 4.1).

It is difficult to prevent the aggregation of fillers, because a diameter of primary particles in reinforcing fillers is of nanometer size. During the rubber processing, rubber engineers have been interested in the technique how to disperse the nanofillers against the strong interactions between the fillers (see Sect. 4.1). The formation of bound rubber by some mechanochemical reactions during mixing filler with rubber is regarded to be one factor to promote the reinforcement effect by filler loading. Chemical modification reactions for the filler surface have been utilized in order to increase the reinforcement effect, by taking advantage of many kinds and grades of commercially available fillers. Uses of various coupling agents [220, 221] have been practiced in order to increase the filler-to-rubber interactions on silica and some other inorganic fillers [83, 84, 222–224]. "Know-how" has been really helpful in practicing rubber compounding, particularly when choosing effective and cheap fillers to a certain rubber system and deciding how to mix them with rubber. Even at present, is it still the case?

(4) Molding and Shaping (Cross-Linking)

The prepared rubber compound is next subjected to cross-linking reactions, including vulcanization. When the product's shape is complex and/or the shape forming is complicated, shaping step is to be considered independently after compounding. In the shape forming step, a few particular properties of rubber, such as tackiness, are important keys of the step which leads to the final good product. Vulcanization is usually conducted in a heated mold, and the final shape of the rubber product is determined by the molding. Thus, the cross-linking process is usually the final stage in the processing except when using TPE.

The rubber compounds usually contain lots of ingredient other than those of cross-linking system, such as a prevulcanization inhibitor (PVI), antioxidants, antiozonants, processing aids (e.g., softener), and fillers, which are more or less reactive and mixed in the processing steps described already. The cross-linking reactions should begin only at the final stage of the processing. In other words, a scorch (as mentioned above, mechanochemically induced chemical reactions, the cross-linking reactions in particular, during the mixing and/or the shaping stage) should be avoided as completely as possible; otherwise, the quality of products becomes poor and often of no use. Therefore, PVI has been developed in order to avoid the worst scorching. In rubber industry, both accelerators and PVI are often mixed together. This is a kind of contradiction, but this unique pair is widely accepted as "a special but indispensable technique" among rubber engineers. In fact, since vulcanization is the most important issue for utilization of rubber, the rubber processing requests highly advanced chemistry for controlling the reactivity of each reagent with rubber and with the other reagents at each step. This is an original base of our regarding rubber processing as reactive processing, involving not only vulcanization (a series of chemical reactions) but also possible mechanochemical reactions during rubber processing. This character as a reactive processing was specifically reflected on the development of so-called reactive processing technique so far mainly applied on polyurethane preparations, which is developing into the reaction injection molding (RIM) technique.

Both curing design (with compounding and cross-linking) and reinforcing design are the most important aspects for the preparation of high-performance rubber products. One of the most serious scientific problems in rubber processing at present is the only insufficiently known reaction mechanism of vulcanization (see Sects. 2.2 and 4.3). However, the present paradigm of vulcanization was established as early as the 1970s, although the details of the reactions involved in vulcanization, which is a complex reaction (see Sect. 2.2), have not been clear enough. Nowadays, the paradigm enables us to design vulcanization step to be conducted systematically under various conditions of each production process. However, the vulcanization paradigm has empirically been established based on the huge number of experimental results (see Sect. 4.3 and Ref. [225, 199). For example, on the basis of the paradigm of vulcanization, one of the most important issues for rubber engineers is a selection of vulcanization system among conventional vulcanization (CV), semi-efficient vulcanization (semi-EV), and efficient vulcanization (EV). Once one of them is selected, preexperiments are started in order to determine the most adequate compounding prescription. Sections 2.2 and 4.3 and textbooks, e.g., Refs. [212–216], are to be consulted for the more details on vulcanization from application viewpoints.

(5) Rubber Processing as a Reactive Processing

As emphasized already, how to control the chemical reactions at each processing step is really important for understanding rubber processing from a scientific point of view. One of the preconditions toward this goal is to know the elementary reactions involved in vulcanization. In other words, it is essential to elucidate the exact mechanism of vulcanization (chemistry of the cross-linking) before the knowledge of the possible chemical reactions at processing steps other than cross-linking. The reason why many engineers, especially mechanical engineers, are not willing to work on rubber is a poor understanding of the rubber chemistry in vulcanization even among rubber chemists. Consequently, establishment of a new technology on "mechanochemistry of rubber" during processing is much anticipated, which should be academically and technically important, together with the elucidation of vulcanization mechanism. Historically, these situations surrounding rubber science have suggested that rubber processing has been pioneering the utilization of mechanochemical reactions, and it now is going to grow into a new arena. Since the molding process with chemical reactions may be called as a reactive molding, the word "reactive processing" is used in this book. However, this technical word is not much familiar yet. Thus, not only in the rubber technology but also in some other fields, the reactive processing is to be recognized as one of the useful techniques to be developed further. For example, reactive injection molding (RIM) mentioned already is establishing itself, which has recently been focused as an important and useful technology in the plastics productions, even though it is yet to be fully established. This unique technology is suggesting that rubber processing is expected to be more generally expanded not only in rubber arena but also in the composite material field in general.

3.5.3 Thermoplastic Elastomer: Elastomer Without Vulcanization?

Thermoplastic elastomer (TPE) [210] is a unique polymeric material, which is molded at high temperature (in terms of processing, exactly like thermoplastics) and shows rubber elasticity at room temperature. Generally, TPE molecule is composed of hard segments and soft segments, and the hard segments aggregate to form hard-segment domains due to thermodynamic incompatibility between hard and soft segments. The size of the hard-segment domains is smaller than the wavelength of visible light, and the resulted higher-order structure formed by the phase separation is named as a microphase-separated structure. A sea-island structure of TPE is a typical morphology to show the rubber elasticity. In other words, the sea shows a continuous rubbery soft phase composed of soft segments, and the islands are the domains of hard segments which play the role of cross-linking sites in the three-dimensional network structure.

As a typical example of TPE, a microphase-separated structure of polystyrene–polybutadiene–polystyrene triblock copolymer (SBS) is shown in Fig. 3.25a [95]. Due to the immiscibility of two segments, S (polystyrene segment) and B (polybutadiene segment), microphase separation occurs in SBS, which results in a network-like structure to give an elastomer. In the temperature dispersion of dynamic modulus of SBS, there observed two transition regions ascribable to the glass transitions of polystyrene (PS) and polybutadiene (PB), and a rubbery plateau region is recognized between the two glass transition temperatures (T_gs). Since the both ends of PB segment are fixed by the hard domains composed of PS segments,

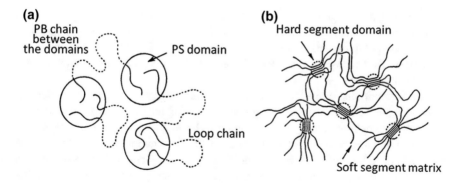

Fig. 3.25 Microphase-separated structures. **a** ABA-type triblock copolymer (A, polystyrene (PS); B, polybutadiene (PB)). **b** Multiblock copolymer (from Fig. 36 in Ref. [95])

the PB segments do not flow at r.t. to function as network chains connecting two hard domains. The observed dynamic modulus of the rubber plateau region is higher than that of PB itself, which is ascribed to a self-reinforcing effect by the presence of PS domains. At a higher temperature region than T_g of PS, the PS domains are melted and the TPE flows like a viscous liquid. This flow enables a melt-molding and recycling of TPE, exactly like thermoplastics. Similar discussion is possible on segmented polyurethane and polyurethaneurea (for the two, see Sect. 3.3.2), whose micro-separated structure is shown in Fig. 3.25b [95]. Figure 3.25 suggests that the three keys for molecular design of high-performance TPE are controls of T_g of (i) a soft-segment matrix, (ii) a dynamic modulus at the rubber plateau region, and (iii) a softening temperature of the hard-segment domains: All of them are to be properly designed in accordance with the properties of the required products.

In terms of the processing and recycling of TPE, the low softening temperature of hard-segment domains are preferable, but the TPE products should be stable up to high temperature for practical applications. For example, a glass transition temperature of PS in SBS is about 100 °C, and the utilization of SBS is limited below about 70 °C. Consequently, in order to increase the thermal stability of the TPE, the use of other polymer of higher melting temperature than PS is necessary. Poly(α-methylstyrene) [226] and poly(ethylene sulfide) [227] were investigated as the candidates of hard segments. However, in the former case the ceiling temperature in the polymerization of α-methylstyrene is low, at around r.t., and a tensile strength of the latter is lower than that of SBS. Thus, poly(alkyl methacrylate) has been focused as an alternative for PS. The high T_g of poly(alkyl methacrylate)s is evident: about 110 °C of poly(t-butyl methacrylate) [228], about 120 °C of syndiotactic poly(methyl methacrylate) (PMMA) [229], about 190 °C of poly(bornyl methacrylate) [230], about 200 °C of poly(isobornyl methacrylate) [231], and about 190 °C of the stereo-complex between syndiotactic PMMA and isotactic PMMA [232]. In addition, poly(1,3-cyclohexadiene) and its hydrogenated one were investigated as hard segments for preparations of thermostable TPE [233].

Furthermore, a hydrogenation reaction has been utilized for the block copolymers in order to improve the thermal stability of the thermoplastic elastomers. For example, polystyrene-poly(ethylene-*co*-butylene)-polystyrene (SEBS) [234, 235] is a typical one, where the hydrogenation of PB segment in SBS was conducted at an industrial manufacturing scale.

Among various kinds of TPE, there are some examples to form a microphase-separated structure by giving a kind of physically cross-linked elastomers: Hydrophilic–hydrophobic difference between soft and hard segments, hydrogen bonding between hard segments, ionic Coulomb interaction between hard segments have been utilized. For examples, poly(oxyethylene) (PEO) segments and D-maltonolactone were introduced to halogenated butyl rubber as graft chains (IIR-*g*-PEO) [236, 237] and as pendant moieties (IIR-*p*-ML) [238] taking advantage of the chemical reactivity of halogen groups, respectively. These amphiphilic polymers (IIR-*g*-PEO, IIR-*p*-ML) were found to be elastomers by the aggregation of each hydrophilic segment in the butyl rubber (hydrophobic) matrix, and the aggregates play a role of physical cross-linking sites. When mesogenic units were chemically introduced to the ends of olefin oligomers, elastomers were obtained by end linking [239].

In order to elucidate the characteristic morphology of TPE, various techniques have been utilized so far, such as three-dimensional transmission microscopy, X-ray scattering, and neutron scattering (see Sect. 4.2.1). Nowadays, the combination of these modern high techniques makes more detailed analysis possible on higher-order structures of TPE, and nanostructural analyses have also become a trend in rubber science.

3.5.4 Dynamic Vulcanizate: Thermoplastic Even with Cross-Linking?

The idea of reactive processing explained in Sect. 3.5.2 was applied not only in the field of plastics like reaction injection molding (RIM) technique, but also in TPE field as described in this subsection. It is a preparation of a thermoplastic vulcanizate (TPV), the processing technique of which was named "dynamic vulcanization" [240–251]. The product TPV shows thermoplasticity even after the dynamic vulcanization. Therefore, it has been significantly focused as a unique rubbery material, and the dynamic vulcanization displays the versatility of rubber processing, which is as described a kind of reactive processing. Also, opposite to the character of TPE which are elastomeric without vulcanization, TPV which is prepared by the dynamic vulcanization maintains thermoplasticity together with rubber elasticity.

Origin of the "dynamic vulcanization" and its process are briefly explained below. The original preparation of TPV is one of the rubber–plastics blending methods, where polypropylene (PP) and cross-linking-type ethylene–propylene

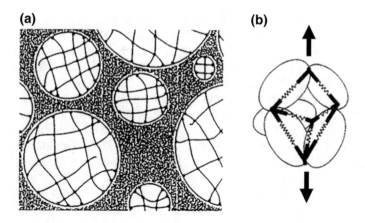

Fig. 3.26 a Morphology of a TPV and **b** Network model of deformed TPV (Modified from Figs. 3 and 4 in Ref. [250])

rubber (EPDM) are used as plastic and rubber components, respectively. Sulfur, organic vulcanization accelerators, stearic acid, and zinc oxide are mixed with the PP-EPDM blend in a mixer under heating above melting point of PP. There, PP is in a melted state, and the mixture of EPDM and the curing reagents are in a dynamic state, i.e., under mechanical mixing. Namely, vulcanization of EPDM took place during the mechanical mixing of PP and EPDM. Note that PP is inert against sulfur, organic accelerators, zinc oxide, and stearic acid. The vulcanization conducted under the mixing is named "dynamic vulcanization," which is in a sense an application of reactive processing of rubber. As a result, the morphology of obtained TPV is speculated as shown in Fig. 3.26 [245]: In the figure (a), particles of vulcanized EPDM are dispersed in a continuous matrix of PP. This morphology is questionable, of course: Why it can be elastic even PP is the matrix which is a typical plastic material? The vulcanized EPDM is rubbery, but it is dispersed as islands in the sea of PP matrix.

Two reasons explaining the rubber elasticity of TPV have been reported [251–253]. The first one is on the elasticity of PP phase located between the vulcanized EPDM particles: The PP was stretchable without yielding even after large deformation, which was explained by stress calculations using a two-dimensional finite element method (FEM) [251, 252]. FEM results showed the calculated stress was lower than the yielding stress of PP. It was proposed that the EPDM prevented the growth of lamella structure of PP by penetrating the EPDM into the PP phase under high shear stress at high temperature. The second explanation has proposed a model to explain the rubber elasticity shown in Fig. 3.26b, where the rubber particles play a role of spring and the continuous phase of PP works as an adhesive for the rubber particles to fix the rubber phases [253].

The exact reasons are still to be revealed, but the dynamic vulcanization is tentatively defined as a technique to prepare a polymer alloy with a morphology shown in Fig.3.26a by mechanical mixing of melted thermoplastics and rubber in the presence of curing reagents, where vulcanization reaction of the rubber component occurs under the influence of shearing stress. Successfully, new rubbery materials have been prepared by using the dynamic vulcanization technique, which is based on the opposite morphology and concept against ordinary TPEs. In them, the traditional cross-linking reactions were eliminated for realizing thermoplastic elastomers. In other words, TPE has established a paradigm shift from "rubber becomes of use by cross-linking" to "even without cross-linking, rubber becomes of use," while TPV is suggesting a new paradigm "even with cross-linking, rubber can be thermoplastic." Therefore, the development of dynamic vulcanization is significant due to the expectation as a new trend in modern rubber science. In fact, various biomasses have been recently used as one component for preparations of new TPV [254]. However, there are still unknown parts in dynamic vulcanization technics, which means we still have to depend on the trial-and-error method. Therefore, it is necessary for rubber scientists to investigate more on the cross-linking reactions under the dynamic conditions, which may possibly related to reactive processing and/or mechnochemistry.

Remark 3 Is Silicone Rubber Inorganic? and Is Fluoro Rubber Organic?
Rubbers are classified into two: One is general-purpose rubber, and the other, special purpose rubber (see **Remark 1**). Silicone rubbers and fluoro rubbers are unique and much different from the traditional rubbers, even among specialty rubbers.

From the chemical structure of silicone rubber chains, $-Si-O-$, it is apparently inorganic. The chemical structure is exactly that of silica and many other inorganic silicate materials. But, is the silicone an inorganic rubber? The answer is both yes and no. In order to be elastomeric, the silicone rubbers have two methyl groups (or ethyl, in anyway, organic group) on Si atom. That is, poly(dimethyl siloxane) (PDMS) is the chemical structure. Its glass transition temperature is the lowest among rubbers, and it may maintain rubber elasticity even in winter Siberia! End-linked PDMS is of use as model network for discussing rubber elasticity (see Sect. 3.2.1). By the way, the "one" in silicone means containing oxygen as in acetone, $(CH_3)_2 C=O$. Therefore, we have to distinguish precisely silicone from silicon. Silicon is inorganic, and silicone rubber is not within silicon polymers.

On the other hand, the main chains of fluoro rubber are composed of $-C-C-$, which suggests fluoro rubber is organic. However, the presence of fluoromethyl group (CF_3-) on carbon atom provides it with anti-oil and anti-chemical properties beyond compare as well as an outstanding weatherability. These distinguished properties are not observed in any organic polymers, and its processing methods to the final products are similar to those of inorganic materials. Hence, fluoro rubber is often considered to be an inorganic rubber [255]. It was originally developed for military uses in USA, and it had been listed by COCOM (Coordinating Committee for Multilateral Export Controls) until its dissolution in 1994. However, even after its breakup, some cases of exporting fluoro rubber products to the former socialist's countries have been discouraged. The performance of the products is so excellent even now.

References

1. H. Hodges, *Technology in the Ancient World* (Penguin Press, London, 1970)
2. G. Basalla, *The Evolution of Technology* (Cambridge University Press, Cambridge, 1988)
3. T.K. Derry, T.I. Williams, *A Short History of Technology: From the Earliest Times to A. D. 1900* (Dover, New York, 1993)
4. L. Beck, *Die Geschichite des Eisens in Technischer und Kulturgeschichtlicher Beziehung, I-V* (Friedrich Vieweg und Sohn, Braunschweig, 1884–1903)
5. G. Agricola, *De Re Metallica*, annotated translation by H. C. Hoover & L. H. Hoover was published in 1912. [Reprint edition of Hoovers' book was published in 1950 by Dover, New York] (1556)
6. Several publications by A. Koyre, E. Zilsel and a few others, that appeared in *J. of History of Ideas* in the early 1940s, took the lead in promoting the idea of the Scientific Revolution
7. H. Butterfield, *The Origins of Modern Science 1300–1800* (G. Bell & Sons, London, 1949)
8. S. Mason, *A History of the Sciences: Main Currents of Scientific Thought* (Abelard-Schuman, London, 1953)
9. J.D. Bernal, *Science in History* (C. A. Waltts & Co., London, 1965)
10. J. Henry, *The Scientific Revolution and the Origins of Modern Science*, 3rd edn. (Palgrave Macmillan, London, 2008)
11. P.M. Harman, *Energy, Force and Matter: The Conceptual Development of Nineteenth-Century Physics* (Cambridge University Press, Cambridge, 1982)
12. B.J. Hunt, *Pursuing Power and Light: Technology and Physics from James Watt to Albert Einstein* (John Hopkins University Press, Baltimore, 2010)
13. J. Mokyr, *The Lever of Riches: Technological Creativity and Economic Progress* (Oxford University Press, New York, 1990)
14. T.J. Misa, *Leonardo to the Internet: Technology & Culture from the Renaissance to the Present* (Johns Hopkins University Press, Baltimore, 2004)
15. T.W. Eagar, in *The Materials Revolution*, ed. by T. Forester (MIT Press, eds., Cambridge, MA, 1988), Chap. 13
16. T.S. Kuhn, *The Copernican Revolution: Planetary Astronomy in the Development of Western Thought* (Harvard University Press, Cambridge, MA, 1957)
17. T.S. Kuhn, *The Structure of Scientific Revolutions*, 2nd edn. (University of Chicago Press, Chicago, 1970)
18. D.C. Lindberg, R.S. Westman (eds.), *Reappraisals of the Scientific Revolution* (Cambridge University Press, Cambridge, 1990)
19. W.H.G. Armytage, *A Social History of Engineering* (Faber and Faber, London, 1961)
20. S. Shapin, *The Scientific Revolution* (University of Chicago Press, Chicago, 1996)
21. R. Kubo, *Gomudansei (Rubber Elasticity)*, (Kawade Shobo, Tokyo, 1947). [In Japanese. Translation of the paragraph in the text into English is done by one of the authors, S. K.]
22. A. Ahagon, Kaut. Gummi Kuntst. **38**, 505 (1985)
23. A. Ahagon, Rubber Chem. Technol. **59**, 187 (1986)
24. A. Ahagon, Rubber Chem. Technol. **66**, 317 (1993)
25. A. Ahagon, Nippon Gomu Kyokaishi **89**, 209 (2016). [In Japanese]
26. K. Urayama, J. Polym. Sci. B, Polym. Phys., **44**, 3440 (2006)
27. K. Urayama, Nippon Gomu Kyokaishi **86**, 94 (2013). [In Japanese]
28. S. Kawabata, H. Kawai, Adv. Polym. Sci. **24**, 89 (1977)
29. L.R.G. Treloar, *The Physics of Rubber Elasticity*, 3rd edn. (Clarendon Press, Oxford, 1975)

30. R.W. Ogden, Rubber Chem. Technol. **59**, 361–383 (1986)
31. T. Kawamura, K. Urayama, S. Kohjiya, Macromolecules **34**, 8252 (2001)
32. K. Urayama, T. Kawamura, S. Kohjiya, Macromolecules **34**, 8261 (2001)
33. T. Kawamura, K. Urayama, S. Kohjiya, J. Polym. Sci., Part B Polym. Phys. **40**, 2780 (2002)
34. K. Urayama, T. Kawamura, S. Kohjiya, J. Chem. Phys. **118**, 5658 (2003)
35. M. Mooney, J. Appl. Phys. **11**, 582 (1940)
36. R.S. Rivlin, D.W. Saunders, Philos. Trans. Roy. Soc. A **243**, 251 (1951)
37. Y. Fukahori, W. Seki, Polymer **33**, 502 (1992)
38. T. Kawamura, K. Urayama, S. Kohjiya, Nippon Reoroji Gakkaishi **31**, 213 (2003)
39. K.J. Smith Jr., A. Greene, A. Ciferri, Kolloid Z. & Z. Polym. **194**, 49 (1964)
40. G. Saccomadi, R.W. Ogden (eds.), *Mechanics and Thermomechanics of Rubberlike Solids* (Springer, Wien, 2004)
41. K. Brüning, *In-situ Structure of Elastomers during Deformation and Fracture* (Springer, Cham, 2014), p. 84
42. M. Fujine, T. Takigawa, K. Urayama, Macromolecules **48**, 3622 (2015)
43. R.A. Stratton, J.D. Ferry, J. Phys. Chem. **67**, 2781 (1963)
44. J.D. Ferry, *Viscoelastic Properties of Polymers*, 3rd edn. (Wiley, New York, 1980)
45. K. Urayama, T. Miki, T. Takigawa, S. Kohjiya, Chem. Mater. **16**, 173 (2004)
46. H. Takino, R. Nakayama, Y. Yamada, S. Kohjiya, T. Matsuo, Rubber Chem. Technol. **70**, 584 (1997)
47. C.M. Roland, *Viscoelastic Behavior of Rubbery Materials* (Oxford University Press, Oxford, 2011)
48. B. Wunderlich, *Thermal Analysis of Polymeric Materials* (Springer, Berlin, 2005)
49. I. Krakovsky, Y. Ikeda, S. Kohjiya, *Thermal Analysis of Rubbers and Rubbery Materials*, ed. by N.R. Choudhury, P.P. De, N.K. Dutta (iSmithers Rapra, Shrewsbury, 2010), Chap. 9
50. S. Hirano, A. Kishimoto, Japanese J. Appl. Phys **3A**(39), 1193 (2000)
51. A. Kato, J. Shimanuki, S. Kohjiya, Y. Ikeda, Rubber Chem. Technol. **79**, 653 (2006)
52. A. Kato, T. Suda, Y. Ikeda, S. Kohjiya, J. Appl. Polym. Sci. **122**, 1300 (2011)
53. A. Kato, Y. Ikeda, S. Kohjiya, *Polymer Composites, Vol, 1: Macro- and Microcomposites*, ed. by S. Thomas, K. Joseph, S.K. Malhotra, K. Goda, M.S. Streekala (WILEY-VCH, Weinheim, 2012), Chap. 17
54. A. Kato, Y. Ikeda, R. Tsushi, Y. Kokubo, N. Kojima, Colloid Polym. Sci. **291**, 2101 (2013)
55. A. Kato, Y. Ikeda, Y. Isono, Nippon Gomu Kyokaishi **87**, 447 (2014). [In Japanese]
56. R.H. Norman, *Conductive Rubbers and Plastics: Their Production, Application and Test Methods* (Applied Science Publishers, London, 1970)
57. V.E. Gul, *Structure and Properties of Conducting Polymer Composites* (VHS, Utrecht, 1996)
58. Y. Wada, *Kobunshi no Denkibussei (Electrical Properties of Polymer)* (Shokabo, Tokyo, 1987). [In Japanese]
59. T. Blythe, D. Bloor, *Electrical Properties of Polymers*, 2nd edn. (Cambridge University Press, Cambridge, 2005)
60. S.A. Abeer, D.E. Reffaee, S.L. Abd-El-Messieh, K.N. Abd-El Nour. Mater. Des. **30**, 3760 (2009)
61. S. Abdul Jawad, A. Alnajjar, Polym. Int. **44**, 208 (1997)
62. K.S. Cole, R.H. Cole, J. Chem. Phys. **9**, 341 (1941)
63. A. Kato, Y. Isono, K. Nagata, A. Asano, Y. Ikeda, *Characterization Tools for Nanoscience & Nanotechnology*, ed. by S.S.R. Kumar Challa (Springer, Berlin, 2014), Chap. 4
64. A. Kato, Y. Ikeda, Nippon Gomu Kyokaishi **87**, 252 (2014). [In Japanese]
65. A. Adachi, T. Kotaka, Macromolecules **18**, 466 (1985)
66. K. Ito (ed.), *Shirikon Handobukku (Hand book of Silicones)* (Nikkan Kogyo Shinbunsha, Tokyo, 1990). [In Japanese]
67. S. Kohjiya, K. Maeda, S. Yamashita, Y. Shibata, J. Mater. Sci. **25**, 3368 (1990)
68. W. Noll, *Chemistry and Technology of Silicones* (Academic Press, New York, 1968)

69. A.G. MacDiarmid (ed.), *Organometallic Compounds of the Group IV Elements,* vol. 1 (Part A. Marcel Dekker, New York, 1968), p. 231
70. S. Kohjiya, A. Ono, S. Yamashita. *Polym.-Plastics Technol. Eng.,* **30**(4), 351 (1991)
71. S. Kohjiya, A. Ono, T. Kishimoto, S. Yamashita, H. Yanase, T. Asada, Mol. Cryst. Liq. Cryst. **185**, 183 (1990)
72. G.W. Gray, P.A. Winsor, *Liquid Crystals and Plastic Crystals* (Wiley, New York, 1974)
73. W. H. de Jeu, ed., *Liquid Crystal Elastomers: Materials and Applications (Adv. Polym. Sci.,* **250**), (Springer, Heidelberg, 2012)
74. S. Kohjiya, T. Hashimoto, S. Yamashita, M. Irie. *Chem. Letters,* 10, 1497 (1985)
75. S. Kohjiya, T. Hashimoto, S. Yamashita, M. Irie, Makromol. Chem., Rapid Commun. **10**, 9 (1989)
76. T. Hashimoto, S. Kohjiya, S. Yamashita, M. Irie. *J. Polym. Sci.: Part A: Polym. Chem.,* **29**, 651 (1991)
77. K. Urayama, Z-h. Luo, T. Kawamura, S. Kohjiya. *Chem. Phys. Letters,* **287**, 342 (1998)
78. M. De Sarkar, K. Urayama, T. Kawamura, S. Kohjiya, Liq. Cryst. 27, 795 (2000)
79. K. Urayama, Y. Okuno, T. Kawamura, S. Kohjiya, Macromolecules **35**, 4567 (2002)
80. K. Urayama, Adv. Polym. Sci. **250**, 119 (2011)
81. K. Urayama, Reac. Func. Polym. **73**, 885 (2013)
82. R. Rauline (To Compagnie Generale des Establissements Michelin). *European Patent* 0501227; *US Patent* 5,227,425 (1993)
83. S. Wolff, Kautsch. Gummi Kunstst. **34**, 280 (1981)
84. A.S. Hashim, B. Azahari, Y. Ikeda, S. Kohjiya, Rubber Chem. Technol. **71**, 289 (1998)
85. G.H. Meeten (ed.), *Optical Properties of Polymers* (Elsevier Applied Science Publishers, London, 1986)
86. A. Kato, Y. Ikeda, Y. Kasahara, J. Shimanuki, T. Suda, T. Hasegawa, H. Sawabe, S. Kohjiya, J. Opt. Soc. Am., Part B **25**, 1602 (2008)
87. A. Kato, Y. Kokubo, R. Tsushi, Y. Ikeda, *Chemistry, Manufacture and Applications of Natural Rubber*, ed. by S Kojiya, Y Ikeda (Woodhead/Elsevier, Cambridge, 2014) (Chapter 9), pp. 193–215
88. A. Kato, Y. Ikeda, Nippon Gomu Kyokaishi **87**, 351 (2014). [In Japanese]
89. K. Matsumura, Y. Kagawa, J. Mater. Sci. Letters **20**, 2101 (2001)
90. K. Miyasaka, *Purasutikku Jiten (Dictionary of Plastics)* (Asakura Shoten, Tokyo, 1992), p. 1024. [In Japanese]
91. S. Wolff, Rubber Chem. Technol. **69**, 325 (1996)
92. M.-J. Wang, Rubber Chem. Technol. **71**, 520 (1998)
93. V.P. Kandidov, V.O. Milisin, A.V. Bykov, A.V. Priezzhev, Quantum Electron. **36**, 1003 (2006)
94. S. Kohjiya, *Natural Rubber: From the Odyssey of the Hevea Tree to the Transportation Age* (Smithers Rapra, Shrewsbury, 2015)
95. S. Kohjiya, *Gomuzairyokagaku Joronn (Introduction to Materials Science of Rubber)* (Nihon Valqua Co., Tokyo, 1995). [In Japanese]
96. Y. Ikeda, in *Solid State Ionics for Batteries,* ed. T. Minami, M. Tatsumisago, M. Wakihara, C. Iwakura, S. Kohjiya, I. Tanaka (Springer, Tokyo, 2005), Chaps. 6.2 and 6.4
97. M. Nagata, Nippon Gomu Kyokaishi **58**, 604 (1985). [In Japanese]
98. S. Kohjiya, Y. Ikeda, S. Yamashita, in *Polyurethanes in Biomedical Engineering II,* ed by H. Plank, I. Syre, M. Dauner, G. Egbers (Elsevier, Amsterdam, 1987), Chap. 14
99. Y. Ikeda, S. Kohjiya, S. Yamashita, H. Fukumura, S. Yoshikawa, Polym. J. **20**, 273 (1988)
100. Y. Ikeda, S. Kohjiya, S. Takesako, S. Yamashita, Biomaterials **11**, 553 (1990)
101. S. Kohjiya, S. Takesako, Y. Ikeda, S. Yamashita, Polym. Bull. **23**, 299 (1990)
102. Y. Ikeda, S. Kohjiya, S. Yamashita, H. Fukumura. *J Mater. Sci.: Materials in Medicine,* **2**, 110 (1991)
103. S. Kohjiya, Y. Ikeda, S. Yamashita, M. Shibayama, T. Kotani, S. Nomura, Polym. J. **23**, 991 (1991)
104. S. Kohjiya, Y. Ikeda, S. Takesako, S. Yamashita. React. Polym. **15**, 165 (1990)

105. Y. Ikeda, M. Tabuchi, Y. Sekiguti, Y. Miyake, S. Kohjiya, Macromol. Chem. Phys. **195**, 3615 (1994)
106. S. Kohjiya, T. Kawabata, K. Maeda, S. Yamashita, Y. Shibata. in *Second International Symposium on Polymer Electrolyte*, ed. by B. Scrosati (Elsevier Applied Science, London, 1990), p. 187
107. S. Kohjiya, T. Horiuchi, S. Yamashita, Electrochim. Acta **37**, 1721 (1992)
108. Y. Ikeda, H. Masui, S. Syoji, T. Sakashita, Y. Matoba, S. Kohjiya, Polym. Inter. **43**, 269 (1997)
109. A. Nishimoto, M. Watanabe, Y. Ikeda, S. Kohjiya, Electrochim. Acta **43**, 1177 (1998)
110. Y. Ikeda, Y. Wada, Y. Matoba, S. Murakami, S. Kohjiya, Electrochim. Acta **45**, 1167 (2000)
111. S. Kohjiya, T. Horiuchi, K. Miura, M. Kitagawa, T. Sakashita, Y. Matoba, Y. Ikeda, Polym. Int. **49**, 197 (2000)
112. Y. Ikeda, Y. Wada, Y. Matoba, S. Murakami, S. Kohjiya, Rubber Chem. Technol. **73**, 720 (2000)
113. Y. Ikeda, T. Kitade, S. Kohjiya, A. Hayashi, A. Matsuda, M. Tatsumisago, T. Minami, Polymer **42**, 7225 (2001)
114. Y. Matoba, Y. Ikeda, S. Kohjiya, Solid State Ionics **147**, 403 (2002)
115. S. Kohjiya, T. Kitade, Y. Ikeda, A. Hayashi, A. Matsuda, M. Tatsumisago, T. Minami, Solid State Ionics **154–155**, 1 (2002)
116. S. Murakami, K. Ueda, T. Kitade, Y. Ikeda, S. Kohjiya, Solid State Ionics **154–155**, 399 (2002)
117. S. Kohjiya, Y. Ikeda, Recent Res. Develop. Electrochem. **4**, 99 (2001)
118. Y.C. Fung, *Biomechanics-Mechanical Properties of Living Tissue-* (Springer, New York, 1981)
119. J.B. Park, *Biomaterials Science and Engineering* (Plenum Press, New York, 1984)
120. N.R. Legge, Rubber Chem. Technol. **62**, 529 (1989)
121. D.J. Lyman, K. Knutson, B. McNeil, Trans. Am. Artif. Int. Organs **21**, 49 (1975)
122. T. Okano, K. Kataoka, *Biomaterial Science*, vol 2 (Nankodo, Tokyo, 1982), pp. 55–70. [In Japanese]
123. F. Croce, G.B. Appetecchi, L. Persi, B. Scrosati. Nature, 394(6692), 456 (1998)
124. Y. Tanaka, Rubber Chem. Technol. **74**, 355 (2001)
125. S. Kohjiya. *Nippon Gomu Kyokaishi*, **88**, 18 & 93 (2015). [In Japanese]
126. Y. Ikeda, A. Tohsan, S. Kohjiya. *Sustainable Development: Processes, Challenges and Prospects*, ed. by D. Reyes (Nova Science Publishers, New York, 2015), Chap. 3
127. S. Kohjiya, Y. Ikeda (eds.), *Chemistry, Manufacture and Applications of Natural Rubber*, ed. by S. Kohjiya, Y. Ikeda (Woodhead/Elsevier, Cambridge, 2014)
128. A.H. Tullo. *Chem. Eng. News*, April 20, 18 (2015)
129. G.G. Lowry (ed.), *Markov Chains and Monte Carlo Calculations in Polymer Science* (Marcel Dekker, New York, 1970)
130. R.-J. Roe, *Methods of X-ray and Neutron Scattering in Polymer Science* (Oxford University Press, New York, 2000)
131. L. Mandelkern, *Crystallization of Polymers*, vols. 1 & 2, 2nd edn. (Cambridge University Press, New York, 2002)
132. S.D. Gehman, Chem. Rev. **26**, 203 (1940)
133. L. Mandelkern, Rubber Chem. Technol. **66**, G61 (1994)
134. F.W. Billmeyer, Jr., *Textbook of Polymer Science*, 3rd edn. (John Wiley & Sons, New York, 1984). [When it was first published in 1957, the title was *Textbook of Polymer Chemistry*]
135. J.R. Katz. Naturwissenschaften, **13**, 410 & 900 (1925)
136. B.N. Zimmerman (ed.), *Vignettes from the International Rubber Science Hall of Fame (1958–1988): 36 Major Contributors to Rubber Science* (Rubber Division, American Chemical Society, Akron, 1989), pp. 47–59
137. S. Murakami, K. Senoo, S. Toki, S. Kohjiya, Polymer **43**, 2117 (2002)
138. S. Toki, I. Sics, S. Ran, I. Liu, B.S. Hsiao, S. Murakami, K. Senoo, S. Kohjiya, Macromolecules **35**, 6578 (2002)

139. M. Tosaka, S. Murakami, S. Poompradub, S. Kohjiya, Y. Ikeda, S. Toki, I. Sics, B.S. Hsiao, Macromolecules **37**, 3299 (2004)
140. M. Tosaka, S. Kohjiya, S. Murakami, S. Poompradub, Y. Ikeda, S. Toki, I. Sics, B.S. Hsiao, Rubber Chem. Technol. **77**, 711 (2004)
141. S. Kohjiya, T. Tosaka, M. Furutani, S. Ikeda, Y. Toki, B.S. Hsiao, Polymer **48**, 3801 (2007)
142. Y. Ikeda, Y. Yasuda, K. Hijikata, M. Tosaka, S. Kohjiya, Macromolecules **41**, 5876 (2008)
143. Y. Ikeda, S. Kohjiya, Nihon Reoroji Gakkaishi **36**, 9 (2008). [In Japanese]
144. Y. Ikeda, Kagaku to Kyoiku (Chemistry and Education) **61**, 472 (2013). [In Japanese]
145. S. Toki, in *Chemistry, Manufacture and Applications of Natural Rubber*, ed. by S. Kohjiya, Y. Ikeda (Woodhead/Elsevier, Cambridge, 2014), Chap. 5
146. M. Yamamoto, J.L. White. J. Polym. Sci., A-2, **9**, 1399 (1971)
147. A. Gros, M. Tosaka, B. Huneau, E. Vernon, S. Pumpradub, K. Senoo, Polymer **76**, 230 (2015)
148. S. Yamazaki, K. Watanabe, K. Okada, K. Yamada, K. Tagashira, A. Toda, M. Hikosaka, Polymer **46**, 1675 (2005)
149. A. Stoddart. Nature Rev. Mats., **2**, 17003. doi:10.1038/naturevmats.2017.3 (2017)
150. S.C. Nyburg, Acta Cryst. **7**, 385 (1954)
151. G.T. Davis, R.K. Eby, J.P. Colson, J. Appl. Phys. **41**, 4316 (1970)
152. G.T. Davis, J.J. Weeks, G.M. Martin, R.K. Eby, J. Appl. Phys. **45**, 4175 (1974)
153. S. Kohjiya, Kagaku Kogyo **38**, 579 (1987). [In Japanese]
154. J. Basti, F. Boué, M. Buzier, in *Molecular Basis of Polymer Networks*, ed. by A. Baumgaertner, C.E. Picot (Springer Verlag, Berlin, 1989), pp. 48–64
155. T. A. Vilgis, F. Boué, S. F. Edwards, in *Molecular Basis of Polymer Networks*, ed. by A. Baumgaertner, C.E. Picot (Springer, Berlin, 1989), pp. 170–177
156. M. Hernandez, M.A. Lopez-Manchado, A. Sanz, A. Nogales, T.A. Ezquerra, Macromolecules **44**, 6574 (2011)
157. T. Karino, Y. Ikeda, Y. Yasuda, S. Kohjiya, M. Shibayama, Biomacromol **8**, 693 (2007)
158. Y. Ikeda, N. Higahsitani, K. Hijikata, Y. Kokubo, Y. Morita, M. Shibayama, N. Osaka, T. Suzuki, H. Endo, S. Kohjiya, Macromolecules **42**, 2741 (2009)
159. T. Suzuki, N. Osaka, H. Endo, M. Shibayama, Y. Ikeda, H. Asai, N. Higashitani, Y. Kokubo, S. Kohjiya, Macromolecules **43**, 1556 (2010)
160. T. Kanaya, Y. Takayama, Y. Ogino, G. Matsuba, K. Nishida, in *Progress in Understanding of Polymer Crystallization*, ed. by G. Reiter, G.R. Strobl (Springer, Berlin, 2007), Chap. 5
161. S. Kohjiya, Y. Ikeda (2017), *Paper presented at the 191st Technical Meeting of Rubber Division, American Chemical Society,* Beachwood, OH, 25–27 April 2017
162. R.B. Merrifield, J. Am. Chem. Soc. **85**, 2149 (1963)
163. L.M. Gierasch, Biopolymers **84**, 433 (2006)
164. S. Trabelsi, P.A. Albouy, J. Rault, Macromolecules **35**, 10054 (2002)
165. S. Trabelsi, P.A. Albouy, J. Rault, Macromolecules **36**, 7624 (2003)
166. S. Trabelsi, P.A. Albouy, J. Rault, Rubber Chem. Technol. **77**, 303 (2004)
167. J.-B. LeCam, B. Huneau, E. Verron, L. Gornet, Macromolecules **37**, 5011 (2004)
168. Y. Miyamoto, H. Yamao, K. Sekimoto, Macromolecules **36**, 6462 (2003)
169. Y. Ikeda, Y. Yasuda, S. Makino, S. Yamamoto, M. Tosaka, K. Senoo, S. Kohjiya, Polymer **48**, 1171 (2007)
170. S. Trabelsi, P.A. Albouy, J. Rault, Macromolecules **36**, 9093 (2003)
171. S. Poompradub, M. Tosaka, S. Kohjiya Y. Ikeda, S. Toki, I. Sics, B.S. Hsio. Chem. Lett. **33**, 220 (2004)
172. S. Poompradub, M. Tosaka, S. Kohjiya, Y. Ikeda, S. Toki, I. Sics, B.S. Hsio, J. Appl. Phys. **97**, 103529 (2005)
173. J. Rault, J. Marchal, P. Judeinstein, P.A. Albouy, Macromolecules **39**, 8356 (2006)
174. S. Toki, I. Sics, B.S. Hsiao, S. Murakami, M. Tosaka, S. Poompradub, S. Kohjiya, Y. Ikeda. *J. Polym. Sci., Part B, Polym. Phys.,* **42**, 956 (2004)
175. S. Toki, I. Sics, S. Ran, L. Liu, B.S. Hsiao, S. Murakami, M. Tosaka, S. Kohjiya, S. Poompradub, Y. Ikeda, A.H. Tsou, Rubber Chem. Technol. **77**, 317 (2004)

176. Y. Ikeda, A. Tohsan, Colloid Polym. Sci. **292**, 567 (2014)
177. K. Brüning, *In-situ Structure of Elastomers during Deformation and Fracture* (Springer, Cham, 2014)
178. Y. Ikeda, P. Junkong, T. Ohashi, T. Phakkeeree, Y. Sakai, A. Tohsan, S. Kohjiya, K. Cornish, RSC Advances **6**, 95610 (2016)
179. P.J. George, C.K. Jacob (eds.), *Natural Rubber: Agromanagement and Crop Processing* (Rubber Research Institute of India, Kottayam, 2000)
180. M.J.W. Cock, M. Kenis, R. Wittenberg, *Biosecurity and Forests: An Introduction* (Forestry Department, Food and Agricultural Organization (FAO) of the United Nations, Rome, 2003)
181. M.R. Finlay, *Growing American Rubber: Strategic Plants and the Politics of National Security* (Rutgers University Press, New Brunswick, 2009)
182. L. Bateman (ed.), *The Chemistry and Physics of Rubber-Like Substances* (MacLaren & Sons, London, 1963)
183. A.D. Roberts (ed.), *Natural Rubber Science and Technology* (Oxford University Press, Oxford, 1988)
184. J. Che, C. Burger, S. Toki, L. Rong, B. Hsiao, S. Amnuaypornsri, J. Sakdapipanich, Macromolecules **46**, 9712 (2013)
185. S. Hashizume. *Nippon Gomu Kyokaishi*, **63**, 71 (1990) [In Japanese]: Extended English version of this article is coauthored by S. Hashizume, Y. Ikeda and S. Kohjiya under the title of *Peculiar Behavior of Natural Rubber in the Mixing Process*, and is to be published from TechnoBiz Communications, Bangkok, in 2018
186. L.A. Wood, N. Bekkedahl, J. National Bur. Stand. **36**, 489 (1946)
187. L.A. Wood, N. Bekkedahl, J. Appl. Phys. **17**, 362 (1946)
188. A. Stevenson, R. Campion. *Engineering with Rubber: How to Design Rubber Compounds*, 2nd edn., ed. A. G. Gent (Hanser, Munich, 2001), p. 192
189. A.M. Cunha, S. Fakirov, *Structure Development during Polymer Processing* (Kluwer Academic Publishers, Dordrecht, 1999)
190. N. Okui, S. Umemoto, R. Kawano, A. Mamun, *Progress in Understanding of Polymer Crystallization*, ed. by G. Reiter, G.R. Strobl (Springer, Berlin, 2007), Chap. 19
191. Y.-K. Lan, A.-C. Su. Polymer, **55**, 3087 (2014)
192. C.W. Bunn, Proc. Roy. Soc. A **180**, 40 (1942)
193. G. Natta, P. Corradini, Angew. Chem. **68**, 615 (1956)
194. Y. Takahashi, T. Kumano, Macromolecules **37**, 4860 (2004)
195. A. Immirizi, C. Tedesco, G. Monako, A.E. Tonelli, Macromolecules **38**, 1223 (2005)
196. G. Rajkuma, J.M. Squire, S. Arnott, Macromolecules **39**, 7004 (2006)
197. A.N. Gent, Trans. Faraday Soc. **50**, 521 (1954)
198. A.N. Gent, Trans. IRI **30**(6), 139 (1954)
199. R. Burfield, Polymer **25**, 1823 (1984)
200. D.E. Roberts, L. Mandelkern, J. Am. Chem. Soc. **82**, 1091 (1960)
201. A. Cameron, W. McGill. J. Polym. Sci.: Part A: Polym. Chem., **27**, 1071 (1989)
202. J.M. Chenal, L. Chazeau, Y. Bomal, C. Gauthier. J. Polym. Sci.: Part B: Polym. Phys., **45**, 955 (2007)
203. S. Murakami, K. Tanno, M. Tsuji, S. Kohjiya. Bull. Inst. Chem. Res., Kyoto Univ., **72**, 418 (1995)
204. Y. Ikeda, A. Kato, S. Kohjiya, S. Takahashi, Y. Nakajima, *Gomukagaku (Rubber Science)* (Asakura Shoten, Tokyo, 2016). [In Japanese]
205. B. Erman, J.E. Mark, *Structures and Properties of Rubberlike Networks* (Oxford University Press, Oxford, 1997)
206. K. Urayama, S. Kohjiya, J. Chem. Phys. **104**, 3352 (1996)
207. K. Urayama, T. Kawamura, S. Kohjiya, J. Chem. Phys. **105**, 4833 (1996)
208. T. Kawamura, K. Urayama, S. Kohjiya, J. Chem. Phys. **112**, 9105 (2000)
209. W.W. Graessley, *Polymer Liquids and Networks: Dynamics and Rheology* (Garland Science, London, 2008)

210. G. Holden, N.R. Legge, R. Quirk, H.E. Schroeder (eds.), *Thermoplastic Elastomers*, 2nd edn. (Hanser Publishers, Munich, 1996)
211. S. Kohjiya, Macromol. Symp. **93**, 27 (1995)
212. S. Kohjiya, *Gomu no Jiten (Encyclopedia of Rubber)* (Asakura Shoten, Tokyo, 2000), pp. 87–96
213. H. Long (ed.), *Basic Compounding and Processing of Rubber* (Rubber Division, American Chemical Society, Akron, 1985)
214. M. Morton (ed.), *Rubber Technology*, 3rd edn. (Chapman & Hall, London, 1995)
215. A.N. Gent (ed.), *Engineering with Rubber: How to Design Rubber Components*, 2nd edn. (Hanser Publishers, Munich, 2001)
216. B. Rodgers, *Rubber Compounding: Chemistry and Applications* (Marcel Dekker, New York, 2004)
217. M. Pike, W.F. Watson, J. Polym. Sci. **9**, 229 (1952)
218. D.J. Angier, W.F. Watson, J. Polym. Sci. **20**, 235 (1956)
219. D.J. Angier, W.T. Chambers, W.F. Watson, J. Polym. Sci. **25**, 129 (1957)
220. E.P. Plueddemann, *Silane Coupling Agents*, 2nd edn. (Plenum Press, New York, 1982)
221. Y. Xie, C.A.S. Hill, Z. Xiao, H. Militz, C. Mai, Composites Part A **41**, 806 (2010)
222. M.P. Wagner, Rubber World **164**, 46 (1971)
223. S. Kohjiya, Y. Ikeda, Rubber Chem. Technol. **73**, 534 (2000)
224. W. Meon, A. Blume, H.-D. Luginsland, S. Uhrlandt, in *Rubber Compounding: Chemistry and Applications,* ed. by B. Rodgers (Marcel Dekker, New York, 2004) (Chapter 7)
225. Y. Ikeda, H. Kobayashi, S. Kohjiya. *Kagaku (Chemistry)*, **70** (6), 19 (2015). [In Japanese]
226. L.J. Fetters, M. Morton, Macromolecules **2**, 453 (1969)
227. M. Morton, S.L. Mikesell. *J. Macromol. Sci., Chem.*, A**7**, 1391 (1993)
228. T.E. Long, A.D. Broske, D.J. Bradley, J.E. McGrath, J. Polym. Sci., Polym. Chem. Ed. **27**, 4001 (1989)
229. J.M. Yu, P. Dubois, P. Teyssie, R. Jerome, Macromolecules **29**, 6090 (1996)
230. A. Matsumoto, K. Mizuta, T. Otsu, J. Polym. Sci., Polym. Chem. Ed. **31**, 2531 (1993)
231. J.M. Yu, P. Dubois, R. Jérôme, Macromolecules **29**, 7316 (1996)
232. E. Schomaker, G. Challa, Macromolecules **21**, 2195 (1988)
233. I. Natori, K. Imaizumi, H. Yamagishi, M. Kazunori. J. Polym. Sci.: Part B: Polym. Phys., **36**, 1657 (1998)
234. W.P. Gergen, R.G. Lutz, S. Davison, in *Thermoplastic Elastomers*, 2nd edn. ed. by G. Holden, N.R. Legge, R. Quirk, H.E. Schuroeder (Hanser, München, 1996), Chap. 11
235. D.R. Paul, in *Thermoplastic Elastomers*, 2nd edn. ed. by G. Holden, N.R. Legge, R. Quirk, H.E. Schuroeder (Hanser, München, 1996). Chap. 15C
236. S. Yamashita, K. Kodama, Y. Ikeda, S. Kohjiya. J. Polym. Sci.: Part A: Polym. Chem., **31**, 2437 (1993)
237. Y. Ikeda, K. Kodama, K. Kajiwara, S. Kohjiya, J. Polym. Sci.: Part B: Polym. Phys., **33**, 387 (1995)
238. Y. Ikeda, Y. Nakamura, K. Kajiwara, S. Kohjiya. J. Polym. Sci.: Part A: Polym. Chem., **33**, 2657 (1995)
239. Y. Ikeda, M. Inaki, A. Kidera, H. Hayashi. J. Polym. Sci.: Part B: Polym. Phys., **38**, 2247 (2000)
240. A.Y. Coran, R.P. Patel, Rubber Chem. Technol. **53**, 141 (1980)
241. A.Y. Coran, R.P. Patel, Rubber Chem. Technol. **53**, 781 (1980)
242. A.Y. Coran, R.P. Patel, Rubber Chem. Technol. **54**, 91 (1981)
243. A.Y. Coran, R.P. Patel, Rubber Chem. Technol. **54**, 892 (1981)
244. A.Y. Coran, R.P. Patel, D. Williams, Rubber Chem. Technol. **55**, 116 (1982)
245. A.Y. Coran, R.P. Patel, D. Williams, Rubber Chem. Technol. **55**, 1063 (1982)
246. A.Y. Coran, R.P. Patel, Rubber Chem. Technol. **56**, 210 (1983)
247. A.Y. Coran, R.P. Patel, Rubber Chem. Technol. **56**, 1045 (1983)
248. A.Y. Coran, R.P. Patel, D. Williams, Rubber Chem. Technol. **58**, 1014 (1985)

249. A.Y. Coran, R.P. Patel, in *Thermoplastic Elastomers*, 2nd edn. ed. by G. Holden, N.R. Legge, R. Quirk, H.E. Schroeder (Hanser, München, 1996), Chap. 7
250. S. Kohjiya, Y. Ikeda. *Shinsozai (New Materials)*, **7**(4), 28 (1996). [In Japanese]
251. Y. Kikuchi, T. Fukui, T. Okada, T. Inoue, Polym. Eng. Sci. **31**, 1029 (1991)
252. T. Inoue, Nippon Gomu Kyokaishi **72**, 514 (1999). [In Japanese]
253. S. Kawabata, S. Kitawaki, H. Arisawa, Y. Yamashita, X. Guo. J. Appl. Polym. Sci.: Appl. Polym. Symp., **50**, 245 (1992)
254. Y. Chen, D. Yuan, C. Xu, ACS Appl. Mater. Interfaces. **6**, 3811 (2014)
255. K. Ihara, S. Kohjiya (eds.), *Fussokei Porima (Fluorine-containing Polymers)* (Kyoritu Shuppan, Tokyo, 1990). [In Japanese]

Chapter 4
Recent Development of Rubber Science

4.1 Reinforcing Nanofillers and Their Aggregation

4.1.1 Rubber/Nanofiller Composite

Carbon black (CB)-loaded rubber vulcanizates which appeared in the early twentieth century [1] and fiber-reinforced plastics (FRPs) which were put into practical use in the latter half of the twentieth century are two typical polymer composites. Tire tread rubber is an example of particulate CB/natural rubber (NR) composite and is the most important part of tires that bears various functions at the interface on road: Here, isotropic properties of rubber are importantly highlighted for "various functions," such as rolling resistance, traction, grip, and wear. Consequently, particulate nanofillers are mostly concerned in this section. The tire tread rubbers (including both the side and the top treads) are further composed with fibers (hence not isotropic) for mechanical reinforcement to form the practical fiber-reinforced CB/rubber composites: The "mechanical" means supporting the heavyweight of car (the car body and persons including loads or freights). Refer to Sect. 5.2 about the structure and functions of pneumatic tires.

Seismic isolation rubber bearings, the structure of which is a lamination of CB-loaded rubber with metallic plates, are explained in **Remark 4**. Its function is to cut off vibration; namely, it is an isolator to protect buildings from earthquake vibration. This specific function is due to rubber not to the plates. In the two cases, the particulate nanofiller-loaded rubber is an isotropic viscoelastic medium, which is effective in displaying various functions as well as mechanical ability to bear loading. The latter is, however, due to the composites with anisotropic fiber or iron plate. Aircraft tires and the rubber bearings support the body of airplane together with lots of crews and passengers, and even the huge buildings, respectively. Therefore, supporting the weight is a static mechanical function which CB/NR composites bear with fibers or metallic plates, as well as various dynamic functions.

© Springer Nature Singapore Pte Ltd. 2018
Y. Ikeda et al., *Rubber Science*, https://doi.org/10.1007/978-981-10-2938-7_4

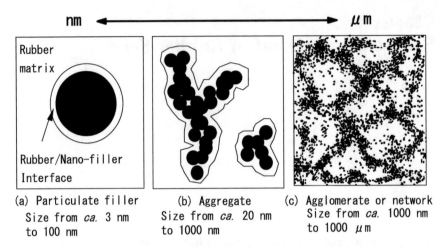

nm ⟵————————————————————⟶ **μm**

(a) Particulate filler
Size from *ca.* 3 nm
to 100 nm

(b) Aggregate
Size from *ca.* 20 nm
to 1000 nm

(c) Agglomerate or network
Size from *ca.* 1000 nm
to 1000 μm

Fig. 4.1 Morphology of nanofiller in rubber matrix (Modified Fig. 1 in Ref. [7])

Rubber is a typical soft material, and it apparently is not tough or tenacious enough to support heavy objects. Yet, mixing of nanofiller has made the soft materials tough enough to support automobiles and buildings in combination with non-rubber components. This amazing effect has long been known as the reinforcement of rubber by nanofillers [2–4], and recently, variously assumed morphology of nanofiller dispersion in the rubbery matrix is under active discussions. For example, it is proposed in a recent review paper [5] that the morphological diagram [5–7] as shown in Fig. 4.1 may give the starting morphological states for the further discussion on the reinforcement of rubber by nanofillers. The morphology shown in the figure is now found to be consistent with the results obtained by three-dimensional transmission electron microscopy (3D-TEM), the details of which are described in the following Sects. 4.1.2 and 4.1.3.

A brief account of Fig. 4.1 is given: **(a)** is a model of particulate nanofiller, e.g., a CB particle (its radius, 3–100 nm). The particle in the rubbery matrix is reasonably assumed to be wetted by encircling rubber layer, which is of low mobility and called bound rubber. The thickness of it is from a few nm to 10 nm, present at the interface between CB and rubber matrix [4–10]. The bound rubber modeled in **(a)** has long history, but it has failed to explain the rubber reinforcement effect quantitatively. Further, an important following fact is not taken into account in this model; that is, CB single particles of nm size produced in the furnace during the manufacturing process have flocculated together to form primary CB aggregates (the size of about 20–1000 nm) due to the van der Waals force (essentially dispersion force for nonpolar molecules and particles) between nm size particles. Therefore, CB does not exist as a single particle as shown in the model **(a)**, but as a primary aggregate. The image of primary CB aggregates may be the ones as shown in **(b)**. When CB is mixed into rubber during the processing, the surface of primary CB aggregates may be covered by rubber layer as schematically shown in the model **(a)** for the assumed primary single particle.

When the CB loading is increased, the size of CB aggregate becomes larger in the mixing process with rubber. That is, the primary CB aggregates further associate together (often called agglomeration) to form agglomerates, even though the surface of them is already covered by bound rubber layer. The detailed structure of the agglomerate has been the target of discussions among rubber scientists and engineers, and this section is focusing on this remaining problem. As one of the possible hypotheses of the CB agglomerate, a rough sketch **(c)** is proposed. The sketch suggests a little network-like association of primary CB aggregates.

Some scientists believed that the CB agglomeration product can possibly be a network structure [2, 4, 10, 11]. However, trials to experimentally show the network structure have failed so far. It has been thought that the pseudonetwork structure is at least one of the reasons to explain the peculiar mechanical properties of nanofiller-loaded rubber vulcanizates such as the Payne effect [12–14] and the Mullins effect [15], and the two effects have been called the "structure" effect of filler. However, it is difficult to completely explain these two effects only by contribution of the filler networks, since the rubber vulcanizate without any fillers may sometimes show a little tendency to the two effects. Recently, the structure was observed by 3D-TEM [16] and was investigated by image analysis techniques [5–8, 16], as described later. Many rubber engineers have believed that both bound rubber and CB networks in rubber vulcanizates are the important factors to explain the reinforcing effect of CB on rubber. Therefore, we could say that three-dimensional observation of the dispersion and aggregation states of nanofiller in rubbery matrix has been a dream of rubber scientists and engineers for a long time. Hereinafter, CB network structure is under consideration as the CB agglomerate.

4.1.2 Experimental of Three-Dimensional Transmission Electron Microscopy

(1) The Microscopic Measurements

Three-dimensional transmission electron microscopy (3D-TEM) is one of the latest structural analytical methods that is also called "electron tomography" [5–7, 16–20], and it enables to visualize the three-dimensional nm-level structures. The instrument used here for 3D-TEM measurements was TECNAI G2 F20 (FEI Co.) [16]. The sample is tilted over a range of angles from $-70°$ to $+70°$, and image data (tilted images) are continuously obtained in $2°$ increments. A total of 71 tilted images are automatically uploaded to the computer. The positions of the tilted images were aligned at that time using the gold colloidal particles. These tilted images are not simply 2D image slices, but rather are 2D-projected images of the mass-density distribution of the samples. This means that 71 consecutive tilted images in total are fed into the computer. Using the Radon transform and the inverse Radon transform installed in the 3D-TEM, the consecutive tilted images obtained in this way are converted to image slices showing the mass-density distribution at each angle. Then, 3D images were reconstructed from the image slices [17, 20]. Please refer to Ref. [17] about the more on the reconstruction method of 3D image.

(2) **Pretreatment of Rubber Samples before Measuring 3D-TEM**

At the first trial of 3D-TEM measurements, all the 3D-TEM slice images of nanofiller-loaded rubber vulcanizate were of very low contrast, and it was very difficult to reconstruct a clear 3D image. Having considered the reason why the contrast of the 3D-TEM image slice was so low, we estimated that the presence of some heavy elements in the sample might have scattered the electron beam to smear the TEM images. The unknown heavy elements, if any, have to be heavier than sulfur, carbon, and hydrogen from which the rubber sample was made. Checking the recipes of the sulfur-cured rubber compounds used in this study, a very likely candidate was found to be Zn atom in zinc oxide. It was mixed in rubber as an activator for vulcanization together with stearic acid. During the vulcanization reaction by sulfur/accelerator (in the present case, a sulfonamide compound), there occurred reactions to generate some unknown rubber-soluble Zn compounds as well as ZnS which is supposed to be the final sulfur product in the vulcanization. Since zinc sulfide is not much soluble in rubber, the more likely candidate is the rubber-soluble Zn compounds, which may be by-products of vulcanization. The possible rubber-soluble Zn compounds may include remaining zinc stearate and a certain zinc complex formed by the reaction of ZnO with a fragment of accelerator. On the base of this estimation, an original pretreatment method (here designated as the NARC-AK method) has been developed for removing soluble zinc compounds from NR vulcanizates by using a special solvent comprised of diethyl ether/benzene/conc. HCl = 43/14/43 (vol%) [20–22]. After the NARC-AK method treatment, the slice images with high contrast were obtained, and 3D image construction was conducted by the 3D-TEM softwares (IMOD [23] and Amira [24]).

4.1.3   Elucidation of Nanofiller Network Structure in Rubber Matrix

(1) **Visualization of Carbon Black Network Structure**

Figure 4.2 presents 3D-TEM images of CB-filled NR vulcanizates after the removal of zinc compounds by the NARC-AK method [8, 25]. The CB loadings of the samples labeled as CB-10, CB-20, CB-40, and CB-80 were 10, 20, 40, and 80 phr CB, respectively. In the figure, the particles which look white are CB, and a black part is the rubber matrix. Even for CB-10 sample with the lowest CB loading, a

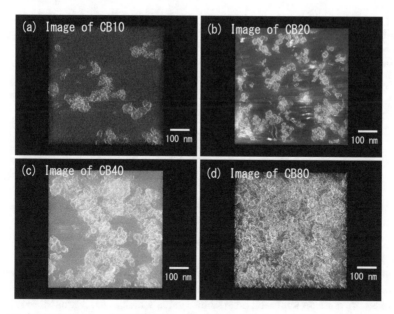

Fig. 4.2 3D-TEM images of CB-filled NR valcanizates after the removal of rubber-soluble zinc compounds (From Fig. 3 in Ref. [8])

number of aggregates are recognized which are composed by several CB particles. It is seen that the CB aggregates increased in size with a higher CB loading. Moreover, for the CB-80 sample with the highest CB loading, it is observed that many large aggregates are positioned closely together. A group of primary particles neighboring within 1-nm distance is defined as CB aggregate. The 1 nm is the resolution limit of the 3D-TEM instrument.

CB has been known to be electron conductive; thus, the results (Fig. 3.6) of volume resistivity showed clear dependency on CB loading amount, and the dependency showed saturation at 30–40 phr of CB loading [5, 7, 8, 25]. Namely, the increase of CB resulted in the increase of conductivity, but it became almost constant around CB of 30 phr, which suggests the CB network formation in the rubber matrix and the percolation resulted in the almost constant highest conductivity of the CB rubber composites. This finding has been reported in many papers on conductivity or resistivity studies of rubber/conducting CB mixtures and interpreted by percolation theory: There observed a threshold value of CB amount to the electron conduction. Reconfirmation of this percolation phenomenon on reinforcing CB in rubber matrix leads us to the further analysis.

From the image analysis of Fig. 4.2, the closest distance between the surfaces of two neighboring aggregates (d_p) was calculated as well as the closest distance between the gravity centers (d_g) [5–8, 25–32]. The dependence of d_p of the CB aggregates on the CB loading (W_{CB}) is shown in Fig. 4.3. It is observed that d_p declined sharply in the W_{CB} region of 20 phr or less, whereas it tended to converge to nearly a constant value in the W_{CB} region of 40 phr or higher. Since it has been

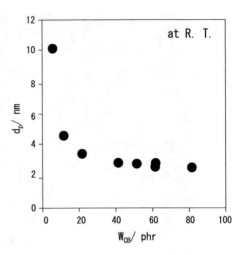

Fig. 4.3 Dependence of the closest distance between the neighboring CB aggregates (d_p) on CB loading (W_{CB}) (From Fig. 7 in Ref. [26])

reported that the bound rubber around CB is between several nm and a few ten nm in thickness [3–5, 8], the result in Fig. 4.3 suggests that at the W_{CB} region of 40 phr or higher, d_p tends to approach one definite value which is ca. 3 nm. Moreover, the dependence (Fig. 3.6) of volume resistivity (ρ_v) on CB loading also indicates the same tendency of the dependence of d_p on CB loading, as mentioned above. The two trends (i.e., the W_{CB} region of 20 phr or less and that of 40 phr or higher in the dependence (Fig. 4.3) of d_p on W_{CB}) suggest that CB aggregates were at a distance of ca. 3 nm from each other in the W_{CB} region of 40 phr or higher, and the distance was the closest CB aggregates were able to approach one another. This value is consistent with the bound rubber thickness (0.6–4.3 nm) estimated by Nishi [33] and O'Brien et al. [34] by pulsed NMR method.

The structure of the CB aggregate networks is visualized by drawing lines connecting the centers of gravity of the closest CB aggregates at a distance of 3 nm, the value at which d_p has saturated at the high CB loading region [5, 7, 25–27, 29, 32]. The resultant network diagrams by this procedure are shown in Fig. 4.4. Diagrams of the 3D network structures in CB-10, -20, -40, and -80 NR vulcanizates are presented. Two types of chains are defined in the CB aggregate networks: cross-linked chains, i.e., network chains which connect CB aggregates, and branch chains which extend outward from the network, having free ends. It is found that cross-linked chains and branch chains of CB aggregates are present in all the samples examined and that there are no isolated chains which are not connected to the CB networks. While isolated smaller networks (maybe called pregels) are observed in the samples having a low W_{CB} of 20 phr or less, no such pregels are found in the vulcanizates with W_{CB} of 40 phr or higher. In other words, in them the CB network structure has connected all CB aggregates, and it is extending all through the sample. Accordingly, the much high conductivity observed in the samples having W_{CB} of 40 phr or higher is attributable to the percolated network where electrons conduct through [19, 35, 36].

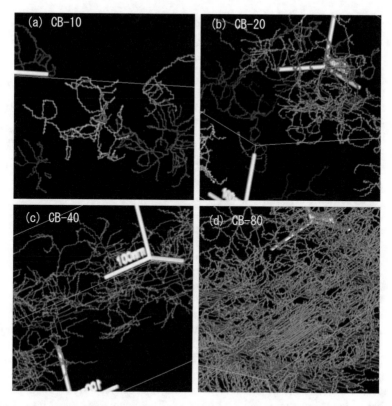

Fig. 4.4 CB network structure of CB-10, -20, -40, and -80 (From Fig. 17.17 in Ref. [29])

(2) Analysis of Filler Networks

Next, the networks of CB are to be analyzed. Figure 4.5 presents a schematic model of the CB network structure having relatively strong interactions, as well as the definitions of the network parameters defined on the model [37, 38]. The circles (O) in the figure represent CB aggregates, and the thin arrows indicate that the aggregates are linked to the surrounding network structures. The two thick arrows in red and in blue represent a cross-linked point and a branched point in the network, respectively. The network parameters are the total volume in the view field of the 3D-TEM image (TV), the number of cross-linked points (N.Nd), the number of branched points (N.Tm), the number of cross-linked chains (N.NdNd), the number of branched chains (N.NdTm), the density of the cross-linked chains (N.NdNd/TV), the density of the branched chains (N.NdTm/TV). Fractal dimension is also calculated as one of the structural parameters.

The fraction of the cross-linked chains (F_{cross}) and the fraction of the branched chains (F_{branch}) are defined by the following equations:

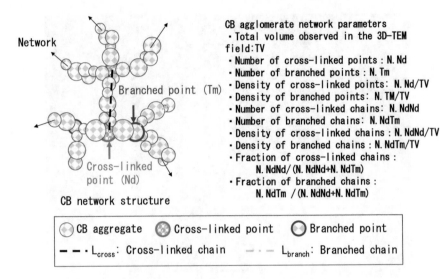

Fig. 4.5 CB network structure and its parameters (From Fig. 6 in Ref. [38])

$$F_{\text{cross}} = N.\text{NdNd}/(N.\text{NdNd} + N.\text{NdTm}) \tag{4.1}$$

$$F_{\text{branch}} = N.\text{NdTm}/(N.\text{NdNd} + N.\text{NdTm}) \tag{4.2}$$

Figure 4.6 shows the dependence of the fraction of cross-linked chains (F_{cross}) and that of branched chains (F_{branch}) on W_{CB}. F_{cross} increases almost linearly with lower W_{CB}, whereas F_{branch} decreases almost linearly in the low W_{CB} at the region

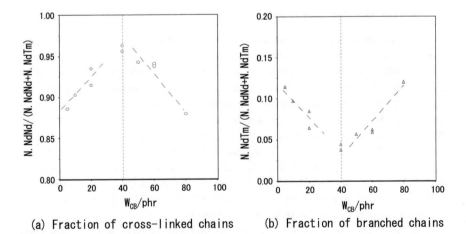

(a) Fraction of cross-linked chains   (b) Fraction of branched chains

Fig. 4.6 Dependence of fraction of cross-linked chains and on CB loading (W_{CB}) (From Fig. 17-20 in Ref. [29])

of less than 40 phr. In contrast, F_{cross} decreases with increasing W_{CB}, whereas F_{branch} increases almost linearly in the high W_{CB} region of 40 phr or more. These results suggest that the network structure in the low W_{CB} region differs from that in the high W_{CB} region after gelation [20, 21, 28–32, 37]. This difference is considered next.

(3) **Process of Filler Network Formation**

Here, it is assumed that the structure of the CB networks formed from CB aggregates is similar to the 3D network structure that is formed by radiation-induced polymerization of monomer (here, primary CB aggregates are treated as monomers) in the presence of a multifunctional monomer (here, the same CB aggregates which are multifunctional in terms of cross-linking point) [37, 38]. In other words, the results are to be explained by the Charlesby's gelation theory [39–41] as follows: The theory treats the formation of soluble polymers (sols) and insoluble polymers (gels). As shown in Fig. 4.5, before the formation of the CB networks, N_y number (range of y is from 1 to n) of CB aggregate fragments of different lengths are present. These fragments contain several numbers of parts ($F \geqq 2$) capable of linking CB aggregates together. The notation F corresponds to the functionality numbers related to the cross-linking reaction of the aforementioned aggregates. The CB aggregate fragments (X_n) are presumed to be composed by the linking of n number of raw CB aggregates of the basic weight unit m. Therefore, a CB aggregate fragment may be regarded as a chain of X_n-mer possessing F.

In addition, this X_n-mer has the distribution of the linked number of the basic weight unit (y, $1 \sim n$). Assuming that there is no linking within the same chain, the percentage of linked $F(q)$ is equal to the probability that one F forms a link, and $(1 - q)$ is the probability that F remains without linking. Considering the total number of the basic weight unit in the view field of 3D-TEM image to be examined, and the total number (N.Nd) of connected points (referred to here as cross-linked points) in the overall field of view can be expressed as shown elsewhere [32], Eq. (4.3) was obtained and indicates that N.Nd/TV is directly proportional to W_{CB} when the formation of the CB network follows the gelation theory [32].

$$N.\text{Nd}/\text{TV} = (q/\xi m\text{TV})W_{CB} \tag{4.3}$$

where W_{CB} is CB loading (phr), ξ is the CB filling correction coefficient (volume of CB filling compound/view field volume of 3D-TEM image) assuming that CB is uniformly dispersed in the rubber, and m is the basic weight unit of raw CB aggregates composing a CB aggregate fragment.

Equation (4.3) indicates that N.Nd/TV is directly proportional to W_{CB} when the formation of the CB network follows the gelation theory. Figure 4.7 shows the density of the cross-linked points (N.Nd/TV) in the CB network as a function of W_{CB}. In the region of $W_{CB} \leqq 20$ phr, the W_{CB} dependence of N.Nd/TV is approximated by the solid straight line (with a slope of $q/\xi m$ TV) passing through the origin in the figure. Consequently, this indicates that the formation of the CB aggregate network in small CB loading region follows the Charlesby's gelation

Fig. 4.7 Relationship
between cross-linking density
(*N*.Nd/TV) of CB network
and W_{CB} (From Fig. 7 in Ref.
[38])

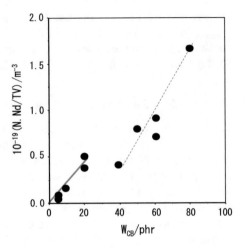

theory [39–41]. (As will be explained later, it should be noted that in the region of
20 phr < W_{CB} < 30 phr, a structural transition occurs with respect to the CB
aggregate structure.) Moreover, also in the region of 30 phr $\leqq W_{CB}$, a linear relationship (dashed line) is observed which has a larger slope than that in the region of
W_{CB} < 20 phr. Presumably, this indicates that *m*, ξ, and TV are not dependent on
W_{CB}. For example, in the region of CB loadings larger than 20 phr, CB network
chains form the agglomerate (pregel) that have larger *F* and *q*, hence the larger
slope. It is inferred that the linking of these chains gives rise to a higher-order
network structure.

4.1.4   Mechanism of Rubber Reinforcement by Nanofillers

In the following discussion on the reinforcement, a hypothesis is included about the
presence of CB/NR interaction layer [21, 25, 30–32]; specifically, constant or
limiting bound rubber layer shown in Fig. 4.3 is being suggestive of this hypothesis. At first, it is reasonably assumed that the interfacial layer between two linked
CB aggregates is equal to or smaller than *ca*.3 nm in thickness as shown in
Fig. 4.3 [5, 25, 26].

   We also have assumed the mixing law. After checking various mixing laws, the
following one is adopted for CB-filled NR vulcanizates, which was somewhat
based on that reported by Facca et al. for plastics [42]. Initially, three components,
NR, CB, and CB/NR interaction layer (CNIL), were taken into accounts, and our
consideration suggests a two-phase mixing model may be enough. In other words,
we have found two components; i.e., NR and CNIL are enough to interpret the
mechanical behaviors. It is no need to say that CB is the most important component
for the rubber reinforcement. However, CB is involved in the formation of CNIL,

and its reinforcing function is possibly assumed to be displayed only through CNIL. Assuming the mixing law of the two phases, the following equations are derived [21, 30–32]:

$$\phi_{ui/s(\text{without CB})} = 100(V_s - V_i - V_{\text{CB}})/(V_s - V_{\text{CB}}) \tag{4.4}$$

$$\phi_{i/s(\text{without CB})} = 100V_i/(V_s - V_{\text{CB}}) \tag{4.5}$$

$$\phi_{ui/s(\text{without CB})} + \phi_{i/s(\text{without CB})} = 1 \tag{4.6}$$

where volumes of CB-filled NR vulcanizate (containing CB), CB, and CNIL (color-coded regions in the figure) are denoted as V_s, V_{CB}, and V_i, respectively. Their respective volume fractions ($\phi_{ui/s(\text{without CB})}$ and $\phi_{i/s(\text{without CB})}$) of rubber matrix and CNIL are defined.

Letting G', G'_{ui}, and G'_i represent the respective moduli of elasticity of the CB-filled NR and NR matrix and the CNIL, the following Eqs. (4.7) and (4.8) are obtained for the two-phase series and parallel mechanical models, respectively.

$$1/G' = \phi_{ui/s(\text{without CB})}(1/G'_{ui}) + \phi_{i/s(\text{without CB})}(1/G'_i) \tag{4.7}$$

$$G' = \phi_{ui/s(\text{without CB})}G'_{ui} + \phi_{i/s(\text{without CB})}G'_i \tag{4.8}$$

Here, visualization and measurement of the CNIL were carried out by image analysis of 3D-TEM images. It is expected that there are some distances (or points) at which the potential energy changes with leaving from the CB/rubber interface. The most stable one of the distances corresponds to the d_p (= $ca.$3 nm) which is obtained from 3D-TEM observation (in Fig. 4.3). It is expected that by connecting CB aggregate clusters by CNIL forms a strong CB network structure. This d_p value is consistent with the thickness value of the glassy state layer around CB, which was found by Nakajima et al. from the AFM results [43].

Figure 4.8 presents 3D images of the CB/NR interaction layer (CNIL) in the CB-10, -20, -40, and -80 samples [31], assuming that the thickness (d_p) of the immobilized rubber interfacial layer was 3 nm. In order to make the CNIL easier to see in this figure, the matrix rubber, CB particles, and CB aggregates are rendered in black, and the CNIL at distances greater than d_p = $ca.$3 nm is shown in different color gradations. The volume fractions of the CNIL and rubber matrix were calculated by image analysis of Fig. 4.8. As a result of the detailed calculation [21, 30, 31], $G'_I = ca.$ 140 MPa was obtained. This G'_i is approximately equal to the modulus of constrained rubber layer around CB, which was reported by Nakajima et al. [43].

From the above discussions, it is understood that a strong CB network is formed by connection of CB aggregates by the mediation of CB/NR interaction layer (CNIL), in which mobility of the rubber chains is constrained. Whereas the size of CB aggregate increases in the W_{CB} region lower than 20 phr, the gelation process is dominating in the W_{CB} region higher than 40 phr in accordance with gelation theory.

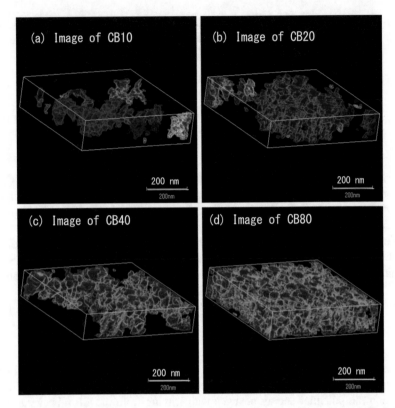

Fig. 4.8 3D image of CB/NR interaction layer in rubber vulcanizate (From Fig. 4 in Ref. [31])

The more details of the reinforcing mechanism are still to be investigated in a near future. For example, the relationship between the volume fraction of the CNIL and the linear expansion coefficient showed a steplike change at about 30 phr CB loading [30, 31]. This phenomenon may suggest a structural phase transition in CB aggregation, which will be worth studying further in relation to the reinforcement effect of particulate nanofillers on rubber.

4.2   Rubber Network Structure Evaluated by Scattering and Spectroscopic Methods

4.2.1   Vulcanization and Heterogeneity of Rubber Networks: Small-Angle X-Ray and Neutron Scatterings

(1) X-Ray and Neutron Beams

The three-dimensional network structures of rubber formed by the vulcanization reactions are of the size between 0.1 nm and a few μm. This size range includes from a few nm up to a few hundred nm range, which has been known as the lost dimension until a recent time. In other words, due to the lack of suitable analytical methods, the structural elucidation of that dimension has been difficult. The small-angle X-ray and neutron scattering methods are now available for covering this gap for structural determination [44, 45].

X-ray is the electromagnetic wave, the wavelength of which is between 0.05 and 0.25 nm. The most used source is a characteristic X-ray of copper (CuKa ray), whose wavelength is 0.154 nm. When the network structure of 1–100 nm scale is to be covered fully by using the X-ray, it is necessary to measure the scattered X-ray at small and ultra-small-angle regions, i.e., from $2\theta = 8.82°$ to $2\theta = 0.0882°$, or in scattering vector (wave vector) from $q = 6.28$ nm^{-1} to $q = 0.0628$ nm^{-1} by the Bragg's condition. The origin of scattering of X-ray is the interaction between incident X-ray and the orbital electrons orbiting around the atomic nucleus, and hence, the scattering results afford structural information of the targeted molecule or sample material. So far, the small-angle X-ray scattering (SAXS) technique has been utilized in morphological studies of thermoplastic elastomers (TPE) rather than network structures of vulcanized rubbers: The main reason may be a smaller difference of electron density between the network rubber chains and the cross-linking sites than that between the phases in TPE [45–47]. Nowadays, however, synchrotron radiation (powerful white X-ray) is available, and a wide range of structural measurements even on elastomeric materials become possible by selecting the monochromatic wavelength of X-ray source.

On the other hand, neutron scattering originated from interaction between neutron and the atomic nuclei allows us to selectively observe a light element, e.g., hydrogen, which is usually difficult to be detected by X-ray scattering [44, 45]. The neutron beam is a particle beam of thermal neutron, i.e., having energy between 0.01 and 0.5 eV, which corresponds to the wavelength of 0.286 and 0.0404 nm, respectively. The interactions between the neutron beam and the nuclei are varied in each atomic nucleus because the involved nuclear forces also changes by atomic nucleus. Thus, an index to show "how many interactions occur per unit time per unit area" by using a microscopic cross-sectional barn (the unit is b), which is defined by $1b = 1 \times 10^{-24}$ cm^2 and is approximately equivalent to atomic nucleus cross section. The neutron scattering contains two types of scatterings, i.e., coherent scattering and incoherent scattering. The coherent scattering is the diffraction from multiple atomic nucleuses, which reflects the structural information. On the other hand, the incoherent scattering is originated from quantum spin of neutron and scattering object. When the spin of scattering nuclei is not zero, the scattering amplitude (the amount relating to the scattering intensity) depends on the direction of spin of neutron, and consequently, the probability of coherent scattering becomes low and the incoherent scattering becomes dominant.

Hydrogen atom has a large incoherent cross section, while deuterium atom (heavy hydrogen) has quite a small incoherent cross section. This difference is used in analyzing the structure of polymer networks, which is of course applicable to rubber networks. When rubber networks are swollen in a deuterated solvent, the contrast between rubber network structure (consisting of carbon and hydrogen atoms) and deuterated solvent (deuterium atoms) is arised. It gives us information on the rubber network structures. This analytical method is named as a swelling–visualization method in small-angle neutron scattering (SANS). Thus, it may be possible to study the network structure of rubber vulcanizates by SANS, using swollen rubber in deuterated toluene. Invisible inhomogeneity in cross-linked rubbers may possibly be visualized by this technique. Note that the swelling and visualization method is, of course, not applicable to all rubber vulcanizates, and applicability may depend on the composition of the rubber vulcanizates: The general principle of applicable or not may remain to be elucidated.

In the following sections, our specific results will be presented as examples applied on elastomers and cross-linked rubbers

(2) Structural Analysis of Rubber Networks by Synchrotron X-ray

The SAXS technique has been applied to ionene elastomers [48, 49], amphiphilic elastomes [50], and organic/inorganic hybrid rubbers [51, 52], as well as TPE in order to characterize their network structures. The SAXS results are generally analyzed using a proposed model structure, which is the important starting point of analysis. For example, the scattering function derived for randomly branched f-functional polycondensates [53], which was introduced from a cascade model [54] based on the Flory–Stockmayer treatment [55], has been very powerful tool for the structural analysis.

On physically cross-linked elastomers, an adequate model and a scattering function are put in accord with the shape of physical cross-linking domains (the aggregated segments), while the cross-linking points are to be considered as the scattering points on chemically cross-linked rubber. The shapes of the domain should be taken into account for the scattering sites, for example, the domains being ion aggregates in the ionene elastomers, assembly of saccharide segments in the amphiphilic elastomers, and silica aggregates in the organic/inorganic hybrid elastomers. By good fittings of the calculated with the observed SAXS data on aliphatic poly(oxytetramethylene) ionene elastomers (IP-Cl and IP-Br), the structural parameters shown in Fig. 4.9 were obtained [50]. The larger ionic domains were formed in the ionene elastomer with chloride counteranions (IP-Cl) than in that with bromide counteranions (IP-Br) to result in the longer distance between the centers of gravity of the domains in the former than the latter. The presence of crystallites in IP-Cl at 25 °C was detected by differential scanning calorimetry and wide-angle X-ray diffraction (WAXD) measurements. The difference in morphology of IP-Cl and IP-Br shown in Fig. 4.9 was well related to that of tensile properties and thermal stability of the network structures: Stress–strain curves and thermal stability of IP-Br were higher than those of IP-Cl, which were ascribable to the larger number of ionic domains in IP-Br.

This SAXS analysis is useful for morphology studies on elastomers which cannot be easily detected by direct observation techniques as well as for in situ analysis of a sol-gel reaction of ethoxysilylterminated oligomers using synchrotron X-ray [51, 52].

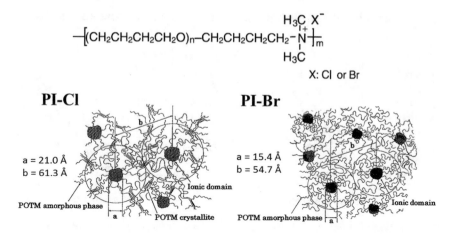

Fig. 4.9 Chemical structures of aliphatic poly(oxytetramethylene) ionenes (IP-Cl and IP-Br) and their schematic morphologies at 25 °C proposed on the basis of SAXS measurements (From Figs. 1 and 7 of Ref. [50])

A comparative study of network inhomogeneity on peroxide cross-linked NR (P-NR) and sulfur cross-linked NR (S-NR) was conducted from a viewpoint of their strain-induced crystallization (SIC, on SIC see Sect. 3.4.2) behaviors. Simultaneous measurements of time-resolved synchrotron WAXD and tensile stress-strain measurements were carried out at SPring-8 [56]. P-NR samples of various network-chain densities (v) were prepared and subjected to the measurements. The stretching ratio at which the crystallization started (α_c) was found to shift to the smaller strain. The higher the v of the sample is, the smaller the α_c became, as shown in Fig. 4.10a. In addition, the values of entropy difference between the undeformed and the deformed states (ΔS_{def}) at α_c were found almost equal in spite of their variation of v. It may be deduced that SIC of P-NR sample begins when the definite ΔS_{def} is reached by stretching regardless of their v, under assumption of same value of entropy at the unstretched state for all P-NR samples [57]. This may be the first experimental results which agree with the prediction by Flory [55].

On the other hand, in S-NR, α_c remained almost same value among all S-NR samples, i.e., α_c was not dependent on v as shown in Fig. 4.10b, while the tendency that the larger was v, the smaller was ΔS_{def} at α_c was maintained [56]. The crystallization rates of both cross-linked samples seemed to increase with increasing v. However, the results showing different α_c suggested that the network structure of P-NR was to be more homogeneous than that of S-NR. Stress dependence of apparent lateral crystallite size was evaluated using the 200 and 120 reflections. From the results, it was found that the variation of L_{200} by nominal stress became smaller in S-NR than in P-NR, which suggested that the stress loaded on the crystallites was smaller in S-NR than in P-NR upon stretching. This finding suggested that there were two phases with high and low network-chain densities in the

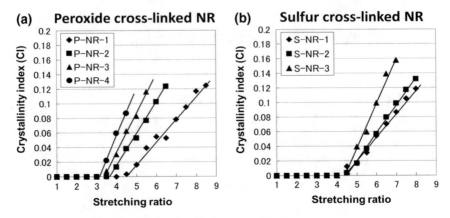

Fig. 4.10 Strain-induced crystallization behavior of **a** peroxide cross-linked NR (P-NR) and **b** sulfur cross-linked NR (S-NR). The larger the number in the sample code, the larger the network-chain density (From Fig. 2 of Ref. [56])

vulcanizate and that the former phase (designated as a network domain) dominantly carried the tensile stress even after the onset of SIC.

From the comparison between SIC behaviors and tensile properties of P-NR, it was found that the larger was v, the larger were both oreinted amorphous index (OAI) and crystalline index (CI) values, which may contribute to the increase of stress. The network domains in rubbery matrix of S-NR were also found to dominantly carry the stress under the stretching before the onset of crystallization. These results may give an answer to the well-known puzzle: Which is a main player to carry the stress in the vulcanizates, the generated crystallites, the oriented amorphous chains, or neither of them? This has been a question in rubber science. In this study, however, it has not fully elucidated why the onset strains are not governed by the network-chain density and SIC starts at almost the same strain. SANS results may also help giving an answer: Almost the same mesh size in the rubbery matrix has resulted in the similar onset strains for the inhomogeneous structure in the sulfur cross-linked isoprene rubber (IR) [58]. More conclusive discussions are still to be pursued.

(3) Analysis of Rubber Network Structure by Small-Angle Neutron Scattering

Invisible homogeneity and inhomogeneity in networks have been quantitatively analyzed by small-angle neutron scattering (SANS) using the swelling–visualization method on polymer gels [44, 45]. This method is applied to sulfur cross-linked isoprene rubbers of various network-chain densities, and the obtained structural parameters of the two-phase inhomogeneous network structures have revealed an unexpected aspect of the vulcanization reaction of rubber [58]. At an early stage of applying SANS measurements on natural rubber (NR), the non-rubber components in NR such as proteins and lipids gave an upturn scattering in the SANS profile, which made the analysis difficult by overlapping with the scattering ascribable to inhomogeneity [59]. Thus, isoprene rubber (IR) is subjected to the SANS method instead of NR. IR compounds were prepared by a conventional mixing of IR, stearic acid (StH), zinc oxide (ZnO), sulfur, and N-cyclohexyl-2-benzothiazole sulfenamide (CBS) on a two-roll mill and were subjected to cross-linking by compression molding at 140 °C for 30 min to obtain IR vulcanizates of 1 mm thickness. Deuterated toluene was used for swelling the samples to be subjected to SANS measurement at the equilibrium of swelling. All SANS profiles clearly showed an upturn scattering in the smaller-angle regions ($q \leq 0.2$ nm^{-1}), suggesting the presence of inhomogeneity in the network structure and supporting the SIC results described in the previous section.

Thus, in this study, the inhomogeneous network structure of swollen IR was speculated to be composed of a two-phase network, where domains of high network-chain density were embedded in rubbery network matrix. In order to analyze the experimentally obtained SANS profiles quantitatively, a series of curve fitting were conducted by using scattering functions similar to the case of swollen gels: In the small-angle region, a squared-Lorentz function (SL-function) was employed to describe solid-like inhomogeneity in the gel, and in the large-angle region, a Lorentz (L) function was used for describing the scattered intensity from a corresponding polymer solution. The characteristic length scale (chord length, Ξ) of solid-like inhomogeneity was considered to be related to the network domains of high network-chain density. The correlation length (ξ) was assumed to show the mesh size in the two-phase inhomogeneous network structure. As a result, the correlation between the recipes for sulfur-curing reagents and the network inhomogeneity was unexpectedly observed.

The size of network domains increased and the mesh size did not changed, even when the concentrations of sulfur and accelerator were increased under the same concentrations of stearic acid and ZnO. It is worth noting that ξ and Ξ linearly decreased more or less by increasing the concentration of ZnO in the region from 0.5 to 2.0 phr in the presence of 2.0 phr stearic acid. When the ZnO concentration was changed in the absence of stearic acid, both ξ and Ξ values were found unchanged, which suggests that stearic acid plays a role to disperse inorganic ZnO into the rubber matrix. Combination and composition of the cross-linking reagents, especially those of zinc oxide with the other reagents, were found to be crucial for the control morphology in the vulcanizates; that is, these results suggest that ZnO plays an important role in combination with stearic acid, in controlling not only the mesh size but also the size of network domains in the cross-linked rubber. In other words, it was found that sulfur cross-linking reagents not only have worked to cross-link rubber chains but also to play a role of controlling the network homogeneity. The inhomogeneous network structure of swollen IR is speculated and illustrated schematically in Fig. 4.11. Additionally, effects of the network inhomogeneity were studied on strain-induced crystallization (SIC) behaviors of the cross-linked IR as revealed by simultaneous time-resolved WAXD and tensile measurements using a synchrotron radiation system at SPring-8. As a result, the mesh size in S-IR was observed to correlate to the onset strain of SIC.

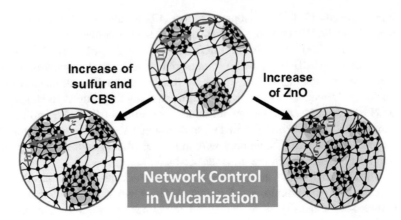

Fig. 4.11 Outline for the control of inhomogeneous network structures in the vulcanizates by sulfur cross-linking reagents (From table content in Ref. [58])

On the basis of these results, it was speculated that the SIC of sulfur cross-linked NR (S-NR) started at similar strains for all S-NR samples, even when the network-chain densities (v) of the network domains were different, and the mesh sizes were almost same regardless of the values of v. The dual use of synchrotron X-ray and neutron scatterings has given us unexpectedly valuable results, which enabled us to clarify a puzzle that has remained unsolved for a long time. In other words, not only the network-chain density (averaged) but also the morphology of inhomogeneous structure in each phase must be considered, when the relationship between the structure and properties would be discussed. The overall network-chain density may not successfully explain the properties of rubber. The neutron scattering is contributing much to the further development of rubber science, together with synchrotron radiation X-ray scattering.

4.2.2 Characterization of Cross-Linked Structure by Spectroscopy

In this section, chemical (the first-order) structure of the cross-linking in the rubber vulcanizates is investigated. X-ray absorption fine structure (XAFS) is an intrinsically quantum mechanical phenomenon: It is based on the X-ray photoelectric effect, in which an X-ray photon incident on an atom within a sample is absorbed and liberates an electron from an inner atomic orbital (e.g., 1 s) [60, 61]. The "photoelectron" wave scatters from the atoms around the X-ray absorbing atom, creating interference between the outgoing and scattered parts of the photoelectron wave function. These quantum interference effects cause an energy-development variation in the X-ray absorption probability, which is proportional to the X-ray absorption coefficient, a measurable quantity. Two regions are recognized in the XAFS: The first is X-ray absorption near-edge structure (XANES), where the peaks, shoulders, and other features near or on the edge are observed, and the second is extended X-ray absorption fine structure (EXAFS) region, which shows the gradual oscillations above the edge. When the XAFS data are decoded properly, these regions provide information about the structure, atomic number, structural disorder, thermal motions of neighboring atoms, and so on.

Nowadays, synchrotron XAFS spectroscopy is one of the most useful techniques enabling in situ XAFS observation on vulcanization reaction of rubber during heating. In order to reveal the details of the two-phase network structure described in the previous section, for example, time-resolved zinc K-edge X-ray absorption fine structure (Zn K-edge XAFS) spectroscopy was conducted in situ during the vulcanization at about 140 °C using synchrotron X-ray at SPring-8 in Japan [62]. The obtained XANES spectra of IR-ZnO-StH-CBS-S_8, IR-ZnSt$_2$-CBS-S_8 and IR-ZnO-CBS-S_8 were decomposed by the linear combo method by using ATHENA in a XAFS analysis package [63]. A good decomposition was obtained in all data points for 50 min as shown in Fig. 4.12. The fitting results clearly suggest that the sulfur cross-linking reaction of IR-ZnO-StH-CBS-S_8 was a combination reaction of two reactions in IR-ZnSt$_2$-CBS-S_8 and IR-ZnO-CBS-S_8 systems. A two-phase network formation via two different reactions was recognized in the sulfur /N-cyclohexyl-2-benzothiazole sulfenamide (accelerator)/zinc oxide/stearic acid curing system, which should be one of the most impact observations in the long history of vulcanization. Furthermore, sulfur K-edge XANES measurement was utilized for the characterization of sulfidic linkages (mono-, di-, and polysulfidic linkages) [64, 65].

Fig. 4.12 Decomposition results of zinc K-edge XANES spectra of IR-ZnO-StH-CBS-S_8 by the linear combo method with reference to zinc K-edge XANES spectra of IR-ZnSt$_2$-CBS-S_8 and IR-ZnO-CBS-S_8. The former and the latter are regarded as the models of mesh network and network domain formations, respectively (From Figs. 1 and 4 in Ref. [62])

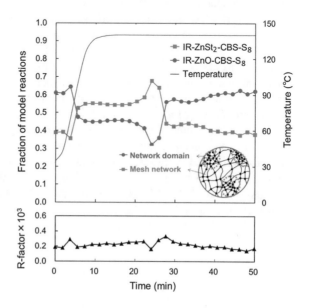

Fig. 4.13 Effect of heat-pressing time on the energies at peak top of sulfur K-edge XANES spectra of the sulfur cross-linked IR samples. The dotted line and two dotted lines indicate the energies at peak edges of polysulfidic and disulfidic linkages of their model samples, respectively (From Figs. 1 and 4 in Ref. [64])

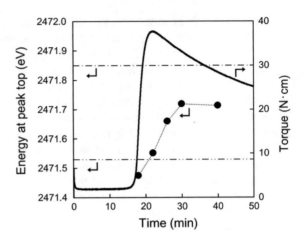

For the CBS-accelerated IR vulcanizates after solvent extractions of unreacted sulfur and accelerator, for example, the sulfidic linkage was found to change from a polysulfidic type to disulfidic type with the increase of the torque under curing as shown in Fig. 4.13 [64]. From the decomposition of the disulfidic linkage, it was also supposed to generate the monosulfidic linkage and monosulfidic type materials where Zn-S linkage is supposed to be generated. The sulfur K-edge XANES technique is easy to be conducted than chemical analyses and will be of use for evaluating a characteristic of the sulfidic linkages in the sulfur cross-linked rubber with the solvent extractions.

Various rubber products have been widely used in our modern society, and all of them are properly cross-linked ones, except adhesives made from liquid rubber solutions and a few thin-film type rubber products prepared from latex. However, the mechanism of vulcanization has not been conclusively clarified yet due to its complex chemical reaction nature (consisting of several elementary reactions, see Sect. 2.2.1 for the complex reaction) between rubber, elemental sulfur, accelerators, activators, and some more additives and reagents annexed for processing and for property improvement of the final products. Further, there have been no quantitative reports on the relations between the inhomogeneity of network structure and properties of cross-linked rubbers, even though many rubber scientists and engineers are now recognizing the rubber vulcanizates. Therefore, the analytical methods introduced in this section would be more in use for elucidating structural inhomogeneity, keeping in mind the ultimate goal of controlling cross-linking reactions and establishing the correlation between the network structure and various properties of final rubber products.

4.3   New Development in Vulcanization Reaction

4.3.1   A New Reactive Intermediate of the Vulcanization Reaction

As mentioned in the previous chapter, powerful tools such as the synchrotron time-resolved X-ray in situ analyses, the SANS measurements, and the related spectroscopic methods have been successfully utilized in the structural studies of rubbers and elastomers. Particularly, in the arena of characterizing the vulcanization (sulfur cross-linking) reaction of rubbers, the observed fundamental results were against a popular belief, to a great surprise of rubber scientists and engineers. This important result is introduced here.

Generally, it has been thought that stearic acid is reacted with zinc oxide (ZnO) to form zinc stearate ($ZnSt_2$, note the ratio of Zn/St = 1/2) in the vulcanization of rubber. Here, St shows sterate. It is a common understanding among all the rubber people so far, and both have been designated as activators for sulfur vulcanization. However, the molar ratio of the zinc ion to stearate of the generated zinc/stearate complex is found to be 1/1, absolutely not 1/2, at 144 °C using in situ time-resolved zinc K-edge XAFS analysis [66, 67]. In Fig. 4.14, the long dotted line shows the concentration of zinc atoms as $(ZnSt)_n^{n+}$. A similar result was obtained even when the amount of ZnO in the IR-ZnO(0.5)-StH(2) compound was decreased to the half

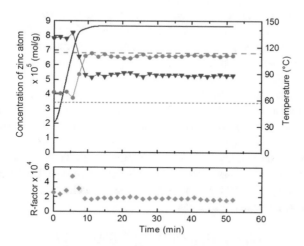

Fig. 4.14 Variation of zinc atoms measured by in situ zinc K-edge XAFS for IR-ZnO(1)-StH(2) under heating. The straight line shows the temperature variation. For the linear combo fittings of IR-ZnO(1)-StH(2), zinc K-edge XANES spectra of $IR-ZnSt_2(4.5)$ (*red filled circle*) and IR-ZnO(1) (*blue down pointing triangle*) under the temperature rising from 35 to 144 °C and those of $IR-ZnSt_2(4.5)$-18.5 and IR-ZnO(1)-18.5 reference samples at 144 °C were utilized as reference spectra, respectively. The R-factor is shown in a mark of (*orange diamond suit*). The long and short dotted lines show the calculated concentrations of zinc atoms as $(ZnSt)_n^{n+}$ and $ZnSt_2$, respectively. The smaller the R-factor is, the better the fitting accuracy is (From Fig. 3a in Ref. [66])

against the concentration of stearic acid to minimize the effect of ZnO in the analyses. Numbers of the sample code show the amount of each reagent in parts per one hundred rubber by weight. Intensity of the white line in zinc K-edge XANES result showed that the coordination number of zinc/stearate complex at 144 °C is four. Additionally, results of in situ Fourier-transform infrared spectroscopy by a single reflection attenuated total reflection (ATR) method suggested the generation of a dinuclear-type bridging bidentate zinc/stearate complex composed of $(Zn_2(\mu\text{-}O_2CC_{17}H_{35})_2)^{2+}\cdot4X$. Here, X is hydroxyl group, water and/or rubber segment.

The skeletal structure of the intermediate is not clear yet, because it is still difficult to conclude whether it is dinuclear, trinuclear, or multinuclear zinc complex, or a mixture of them. However, a trend that a four-coordinated zinc-oxygen metal-organic framework is unstable in water was reported [68]. Also, it is well known that the dinuclear zinc complexes play an important role in enzymatic reactions in our bodies [69]. Therefore, formation of a dinuclear-type structure is proposed to be most likely for the zinc/stearate complex during vulcanization under the presence of water. This proposal is supported by two characteristic features of vulcanization: The first is the results of SANS study on the inhomogeneity of the vulcanizates [58], where the mesh sizes obtained by SANS were almost same among the samples prepared by using the same amounts of zinc oxide and stearic acid as described in Sect. 4.2.1. The inhomogeneous dispersion of curing reagents in a rubber matrix has been considered as one of the serious problems in rubber industry due to the mechanical mixing in the solid state, which is possibly much different from the solution mixing, even though the resultant averaged mesh sizes were not much different. The second was the well-known phenomenon in rubber science that the network formation under the presence of zinc oxide and stearic acid is generally very fast (hence, both are activators). The estimated enzyme mimic reaction intermediate in vulcanization must have accelerated sulfur cross-linking reaction of rubber.

The density functional theory (DFT) using the Gaussian 09 program [70] supported the structures of the speculated intermediates by comparing wave numbers between experimental and theoretical infrared spectra. Figure 4.15 shows experimental and calculated infrared spectra in the range between 1700 and 1300 cm^{-1} to show the carboxylate antisymmetric and symmetric peaks and their shifts. These bands were mostly in a good agreement with the experimental infrared spectra of IR-ZnO(0.5)-StH(2) at 144 °C. Therefore, the structure of the most probable intermediates is generally formulated as

$$\left(Zn_2(\mu\text{-}O_2CC_{17}H_{35})_2\right)^{2+}(OH^-)_2\cdot XY$$

where X and Y are water and/or a rubber segment.

Quantum mechanical approach on the electronic structures has been restricted to small molecules, since Shrödinger equation became the master equation of atoms, molecules, and crystalline systems. As shown in the study described above, however, great progress in computer hardware and software during a few decades

Fig. 4.15 Experimental and calculated infrared spectra in the range between 1700 and 1300 cm^{-1} to show the carboxylate antisymmetric and symmetric bands and their shifts. The experimental spectra of IR-ZnO(0.5)-StH (2) at 144 °C. The carboxyl shift was utilized to identify the kind of substituent on the complex. Here, R in the structure is a rubber segment (From Fig. 8 in Ref. [66])

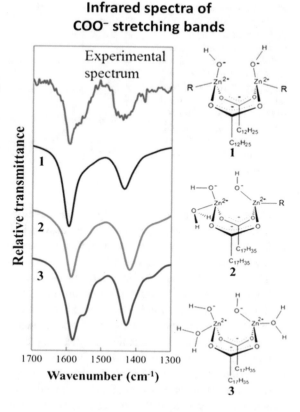

opened a new horizon of theoretical calculations, toward larger and more complex systems [67]. Since the 1990s, the accuracy of DFT methods has reached the chemical accuracy owing to improvement of numerical integration technique and development of new functional forms [71, 72]. Nowadays, researchers can perform the highest level calculations on personal computers, which are faster than workstations of ten years ago and faster than mainframe computers of thirty years ago. Beyond the progress in hardware, computational algorithm has also advanced in several aspects. Invention of the energy-gradient method made optimization of the local minima and characterization of the transition states remarkably facile and routine work. For treatment of heavy atoms, the development and implementation model-core potential method made calculations including several transition metal atoms possible. Thus, the target of theoretical calculations has extended from the world of angstrom to that of several nanometers. Next aspect of amazing development is computer graphics [70]. This technique is not related to the quantum theory. However, the visualization of electron density, orbital phase, and electric potential helps researchers' understanding to be very facile and to accelerates next action in computational work (submitting a job). The visualization of vibrational normal modes is also very helpful to comprehend the type of vibration and find the

Fig. 4.16 Cure curve of IR-ZnO(0.5)-StH(2)-CBS(1)-S$_8$(1.5) measured by using a parallel plate-type rheometer (MR-500, Rheology Co., Kyoto) at 1 Hz of the frequency. Temperature was set according to a controlled program from 35 to 144 °C for about 10 min and at 144 °C for a definite time (From Supporting information of Ref. [66])

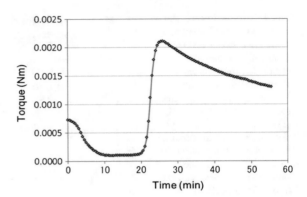

characteristic normal mode. Owing to the developments mentioned, theoretical approach becomes more popular technique even in the field of rubber chemistry, and the comparison between calculation and experiments becomes more important in several aspects of the research [67].

The cure curve shown in Fig. 4.16 suggests an adequate curing behavior of the compound for rubber productions, because the compounds are required to be flowed to the ends of the mold in order to produce the desired shaped rubber product. In the CBS vulcanization system under the presence of zinc oxide and stearic acid, the newly observed zinc/stearate complex may play a role to generate an active intermediate to accelerate the sulfur cross-linking reaction of rubber probably via enzyme mimic reactions after the adequate waiting time (scorch time). This reaction mechanism may have been predicted by nobody of rubber scientists and technologists. It is very unique that the establishment of vulcanization by many rubber researches for last a few decades has an interesting relationship with the evolution of living bodies since tens of thousands of years.

Then, how the intermediate accelerates the vulcanization reaction of rubber? The detailed discussion on the whole vulcanization mechanism will be reported in a near future. Time-resolved measurements and theoretical chemistry will bring about the final goal to reveal it. The results will be expected to be of use to produce valuable rubber products for constructing the safe and sustainability society on the Earth.

4.3.2   New Paradigm of Vulcanization in the Twenty-First Century?

Long history of the vulcanization technique until 2015 (when Ref. [66] was published) is summarized, starting from the innovation by Goodyear in Sect. 2.3.1, followed by the descriptions in Sects. 2.3.2 and 2.3.3 about the sophisticated

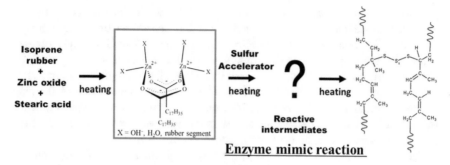

Fig. 4.17 Predicted reactions of the vulcanization system in which zinc oxide, stearic acid, and CBS are added to the rubber matrix (From Fig. 4 in Ref. [67])

vulcanization technique established in the 1960s or 1970s. (It is assumed that the first "paradigm" of the vulcanization technique was established then. Refer to Sect. 2.3.1 and Ref. [73] about paradigm.) In the previous section, a new reactive intermediate in the elementary reaction of vulcanization is introduced, which was reported in 2015 [66, 67]. From the viewpoint of elementary reactions in the mechanism of vulcanization, the meaning of the newly observed intermediate is discussed, and a possibility of the second paradigm of vulcanization technique in the twenty-first century is suggested in this subsection.

The intermediate was generated in the rubber matrix via a reaction of zinc oxide and stearic acid during heating at the vulcanization temperature. See the first step of the reaction scheme shown in Fig. 4.17. Contrary to the traditional understanding that one zinc ion (Zn^{2+}) combines with two stearates, the ratio of zinc ion to stearate in the proposed intermediate is 2/2. The traditionally assumed zinc/stearate complex, whose ratio is 1/2, is stable and is dissolved in a nonpolar rubber matrix. Therefore, it has been reasonably thought to be an important complex for vulcanization of rubber. According to this idea, the vulcanization reaction must have proceeded more smoothly when the isolated zinc/stearate complex of the ratio of 1/2 is added. However, trials of adding the 1/2 complex to the system, instead of mixing the two components separately, have failed to show any improvement in promoting the vulcanization, and this method has not been used in rubber industry. From a viewpoint of reaction kinetics, it can be concluded that the formation of the zinc/stearate complexes with the ratio of 1/2 is not the rate-determining step of the vulcanization in the rubber matrix.

The rate-determining step of vulcanization reaction is, therefore, some other elementary reaction to be identified still now. The vulcanization reaction is a complex reaction in which several elementary reactions are involved. In fact, the vulcanization system includes a lot of reagents or chemicals, and they may interact with each other variously. (Refer to Sect. 2.2.1(**1**) about complex and elementary reactions.) Chemical kinetics of a complex reaction shows that the rate of the overall reaction is determined by the slowest elementary reaction. On the base of this rate-determining step concept, finding of the new zinc/stearate complex with the zinc/stearate ratio of 2/2 does not reject the generation of the traditionally thought Zn/St complexes with the ratio of 1/2, of course. Kinetics simply suggests that its formation is not a rate-determining step in the vulcanization. At present, the formation of the new complex can be a candidate of the rate-determining step to determine the overall vulcanization rate. This shows the importance of the zinc/ stearate complexes with the ratio of 2/2 from the viewpoints of chemical kinetics. More details of the possible elementary reactions involved are to be elucidated in a near future for the development of vulcanization mechanism.

Probably, finding of dinuclear-type bridging bidentate zinc/stearate complex composed of $(Zn_2(\mu\text{-}O_2CC_{17}H_{35})_2)^{2+}\cdot 4X$ (the zinc/stearate complexes with the ratio of 2/2) may be an important starting step to elucidate the mechanism of vulcanization, since the reactive intermediate may be assumed to accelerate the vulcanization reaction more than the traditional $Zn(St)_2$ complex, which is relatively stable or itself is not much reactive. For the last few decades, the studies on the "chemical mechanism of vulcanization" have been active in order to open a new window of rubber vulcanization system [67, 74–80]. Yet, the mechanism has not been satisfactorily revealed, and rubber scientists are obliged to develop the vulcanization design for higher-performance rubber products on the basis of their past experiences depending on the trials and errors method. The finding of new active intermediate may open a door toward new stage of vulcanization, which is still an indispensable technology in rubber science. It may give rise to a "breakthrough" in rubber technology, possibly leading to the new paradigm of vulcanization. Figure 4.17 shows our estimation on the rubber vulcanization technology, on an occasion of the discovery of new reactive intermediate [67, 81].

For a moment, the exact function of the found complex is to be elucidated more. For the vulcanization of rubber, both X and Y or at least either X or Y in the complex has to be a rubber segment. One possible assumption to be examined is a function of water, which may occupy the ligand sites X and Y. This is to be tested in a near future, but a preliminary account of the function of water is given here. The ligand field of the zinc/stearate complexes of 2/2 is in the atmosphere of relatively nonpolar rubber matrix. Around the complexes, however, there exists a small amount of water due to the atmospheric moisture. The complex itself is relatively of higher polarity in the rubber matrix, and hence, water molecules are the probable ligands of the complex at r.t., if any. However, rubber segments are estimated to predominantly occupy the ligand sites of the complex at the vulcanization temperature (about 140 °C in our experiments, and over 180 °C in the case of high-temperature vulcanization, which is more probable in rubber industries for high-speed vulcanization). Due to the high temperature, the participation of moisture may be neglected. Also, it is notable that at high temperatures, the mobility of rubber segments is large enough to move to the ligand sites and to leave them for exchanging the site. Therefore, it is reasonably assumed that both X and Y or at least either of them is a segment of rubber molecules. Further involvements of accelerators and some other species participating in the vulcanization reactions are now under investigation.

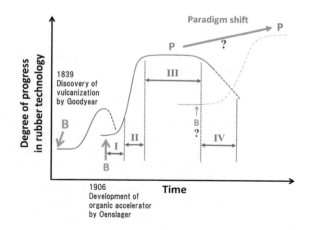

Fig. 4.18 Historical developmental pattern of vulcanization technique of rubber. I the period of breakthroughs; II the active developmental period; III the period of maturity; IV the period of decline; B breakthrough; P paradigm: ⇒, paradigm shift (From Fig. 4.17 in Ref. [81])

From a historical viewpoint of technology development, in particular that of the vulcanization since the invention of it by C. Goodyear, the discussions described above are mapped as shown in Fig. 4.18. The invention of vulcanization by Goodyear was the first breakthrough, and it was the beginning of the development of rubber vulcanization. The followed numerous studies on improving the initial technique of vulcanization using sulfur and lead white had narrowly developed to sustain the mass production of cars early in the twentieth century, particularly in the USA. Ford Model T was on the market in 1908.

The invention of organic vulcanization accelerators was the second breakthrough and finally enabled tires of cars to be produced just-in-time. Just after G. Oenslager, the period of developmental stage (I) is recognized, when a lot of organic accelerators were proposed, followed by the developmental stage (II) when various organic accelerators were industrially used. During the 1960s and the 1970s, the techniques of rubber vulcanization were widely used and matured, which was the period of maturity (III). Here, we may recognize that the first paradigm of vulcanization was established. Then, the maturity stage is destined to decline, i.e., the stage (IV) should come sooner or later. About 50 years have passed since the establishment of the vulcanization paradigm, many rubber-related people may be enjoying the maturity, but some may be looking for a new discovery. The finding of a new reaction intermediate is at the initial stage of decline, or it may mark the final stage (IV) is a question to ask now.

If we can assume the present discovery of new reactive intermediate is at almost the final stage of (IV), it may be found to be another breakthrough or the third one on vulcanization. If this is the case, a period of many trials relevant to the reactive species of vulcanization reactions may follow, ultimately leading to the 2nd paradigm of vulcanization in this century. Anyway, however, it is obvious that modern approach is indispensable for further clarifying the complicated mechanism of vulcanization reactions. Readers are requested to study more on the challenging problems in the vulcanization arena.

References

1. B.N. Zimmerman (ed.), *Vignettes from the International Rubber Science Hall of Fame (1958-1988)* (Rubber Division of Am. Chem. Soc, Akron, 1989), pp. 193–197
2. G. Kraus (ed.), *Reinforcement of Elastomers* (Interscience Publishers, New York, 1965)
3. G. Kraus, in *Science and Technology of Rubber*, ed. by F.R. Eirich (Academic Press, New York, 1978), Chap. 8
4. J.B. Donnet, R.C. Bansal, M.J. Wang (eds.), *Carbon Black* (Marcel Dekker, New York, 1993)
5. S. Kohjiya, A. Kato, Y. Ikeda, Prog. Polym. Sci. **33**, 979 (2008)
6. S. Kohjiya, A. Kato, J. Shimanuki, T. Hasegawa, Y. Ikeda, Polymer **46**, 4440 (2005)
7. S. Kohjiya, A. Kato, T. Suda, J. Shimanuki, Y. Ikeda, Polymer **47**, 3298 (2006)
8. A. Kato, J. Shimanuki, S. Kohjiya, Y. Ikeda, Rubber Chem. Technol. **79**, 653 (2006)
9. P.B. Stickney, R.D. Falb, Rubber Chem. Technol. **37**, 1299 (1964)
10. G. Kraus, Rubber Chem. Technol. **38**, 1070 (1965)
11. F. Bueche, *Physical Properties of Polymers* (Interscience Publishers, New York, 1962)
12. A.R. Payne, J. Appl. Polym. Sci. **6**, 57 (1962)
13. A.R. Payne, J. Appl. Polym. Sci. **6**, 368 (1962)
14. A.R. Payne, R.E. Whittaker, Rubber Chem. Technol. **44**, 440 (1971)
15. L. Mullins, Rubber Chem. Technol. **42**, 339 (1969)
16. Y. Ikeda, A. Kato, J. Shimanuki, S. Kohjiya, Macromol. Rapid Commun. **25**, 1186 (2004)
17. J. Frank (ed.), *Electron Tomography: Three-Dimensional Imaging with the Transmission Electron Microscope* (Plenum Press, New York, 1992)
18. K.P. Dejong, A.J. Koster, ChemPhysChem **3**, 776 (2002)
19. M. Weyland, P.A. Midgley, Mater. Today **7**, 32 (2004)
20. A. Kato, Y. Ikeda, Y. Sato, E. Nagano, Nippon Gomu Kyokaishi **87**, 203 (2014) [In Japanese]
21. A. Kato, Y. Isono, K. Nagata, A. Asano, Y. Ikeda, in *Characterization Tools for Nanoscience & Nanotechnology*, ed. by S.S.R.K. Challa (Springer, Berlin, 2014), pp. 140–189
22. A. Kato, Kobunshi **63**, 632 (2014) [In Japanese]
23. J.R. Kremer, D.N. Mastronarde, J.R. McIntosh, J. Struct. Biol. **116**, 71 (1996)
24. S. Detlev, W. Malte, H. Hans-Christian, in *The Visualization Handbook*, ed. by C.D. Hansen, C.R. Johnson (Elsevier, Amsterdam, 2005), pp. 749–767
25. S. Kohjiya, A. Kato, J. Shimanuki, T. Hasegawa, Y. Ikeda, J. Mater. Sci. **40**, 2553 (2005)
26. S. Kohjiya, A. Kato, Kobunshi Ronbunshu **62**, 467 (2005) [In Japanese]
27. Y. Ikeda, A. Kato, J. Shimanuki, S. Kohjiya, M. Tosaka, S. Poompradub, S. Toki, B.S. Hsiao, Rubber Chem. Technol. **80**, 251 (2007)
28. A. Kato, S. Kohjiya, Y. Ikeda, Rubber Chem. Technol. **80**, 690 (2007)
29. A. Kato, Y. Ikeda, S. Kohjiya, in *Polymer Composites Volume 1: Macro- and Microcomposites*, ed. by S. Thomas, K. Joseph, S.K. Malhotra, K. Goda, K.M.S. Streekala (Wiley-Vch, Boscher, 2012), pp. 515–543
30. A. Kato, Y. Ikeda, R. Tsushi, Y. Kokubo, J. Appl. Polym. Sci. **130**, 2594 (2013)
31. A. Kato, Y. Ikeda, R. Tsushi, Y. Kokubo, N. Kojima, Colloid Polym. Sci. **291**, 2101 (2013)
32. A. Kato, Y. Ikeda, S. Kohjiya, Nippon Gomu Kyokaishi. **88**, 3 (2015) [In Japanese]
33. T. Nishi, J. Polym. Sci., Polym. Phys. Ed. **12**, 685 (1974)
34. J. O'Brien, E. Cassshell, G.E. Wardell, V.J. McBriety, Macromolecules **9**, 653 (1976)
35. A. Kato, Y. Isono, J. Appl. Polym. Sci. **128**, 2498 (2013)
36. A. Kato, Y. Ikeda, Y. Isono, Nippon Gomu Kyokaishi **87**, 447 (2014) [In Japanese]
37. A. Kato, Y. Ikeda, Network Polymer **33**, 267 (2012) [In Japanese]
38. A. Kato, Y. Ikeda, Nippon Gomu Kyokaishi **87**, 252 (2014) [In Japanese]
39. A. Charlesby, J. Polym. Sci., Part B: Polym. Phys. **11**, 513 (1953)
40. A. Charlesby, Proc. Roy. Soc. A **222**, 542 (1954)
41. A. Charlesby, J. Polym. Sci., Part B: Polym. Phys. **14**, 547 (1954)
42. A.G. Facca, M.T. Kortschot, N. Yan, Composites. Part A: Appl. Sci. Manuf. **37**, 1660 (2006)

43. K. Nakajima, T. Nishi, in *Current Topics in Elastomers Research*, ed. by A.K. Bhomik (CRC Press, Boca Raton, 2008), pp. 579–604
44. M. Shibayama, Macromol. Chem. Phys. **199**, 1 (1998)
45. T. Zemb, P. Lindner, *Neutrons, X-rays and Light: Scattering Methods Applied to Soft Condensed Matter* (Elsevier, Amsterdam, 2002)
46. B.P. Grady, S.L. Cooper, in *Thermoplastic Elastomers, Science and Technology of Rubber*, 2nd edn., ed. by J.E. Mark, B. Erman, F.R. Eirich (Academic Press, San Diego, 1994), Chap. 13
47. G. Holden, N.R. Legge, R.P. Quirk, H.E. Schroeder, *Thermoplastic Elastomers*, 2nd edn. (Hanser, Munich, 1996)
48. Y. Ikeda, T. Murakami, Y. Yuguchi, K. Kajiwara, Macromolecules **31**, 1246 (1998)
49. Y. Ikeda, T. Murakami, K. Kajiwara, J. Macromol. Sci. –Phys. **B40**, 171 (2001)
50. Y. Ikeda, J. Yamato, T. Murakami, K. Kajiwara, Polymer **45**, 8367 (2004)
51. H. Urakawa, Y. Ikeda, Y. Yuguchi, K. Kajiwara, Y. Hirata, S. Kohjiya, in *Polymer Gels and Networks,* ed. by Y. Osada, A.R. Khokhlov (Marcel Dekker, New York, 2001), Chap. 1
52. H. Urakawa, Y. Yuguchi, Y. Ikeda, K. Kajiwara, Y. Hirata, S. Kohjiya, in *Polymer Gels Fundamentals and Applications,* ed. by H.B. Bohidar, P. Dubin, Y. Osada (ACS, Washington, DC, 2002), pp. 70–79
53. K. Kajiwara, W. Burchard, M. Gordon, Brit. Polym. J. **2**, 110 (1970)
54. I.L. Good, Proc. Cambridge Phil. Soc. **51**, 240 (1955)
55. P.J. Flory, *Principles of Polymer Chemistry* (Cornell University Press, Ithaca, 1956)
56. Y. Ikeda, Y. Yasuda, K. Hijikata, M. Tosaka, S. Kohjiya, Macromolecules **41**, 5876 (2008)
57. Y. Ikeda, Y. Yasuda, S. Makino, S. Yamamoto, M. Tosaka, K. Senoo, S. Kohjiya, Polymer **48**, 1171 (2007)
58. Y. Ikeda, N. Higashitani, K. Hijikata, Y. Kokubo, Y. Morita, M. Shibayama, N. Osaka, T. Suzuki, H. Endo, S. Kohjiya, Macromolecules **42**, 2741 (2009)
59. T. Karino, Y. Ikeda, Y. Yasuda, S. Kohjiya, M. Shibayama, Biomacromolecules **8**, 693 (2007)
60. G. Bunker, *Introduction to XAFS* (Cambridge University Press, Cambridge, 2010)
61. S. Calvin, *XAFS for Everyone* (CRC Press, Boca Raton, 2013)
62. Y. Yasuda, S. Minoda, T. Ohashi, H. Yokohama, Y. Ikeda, Macromol. Chem. Phys. **215**, 971 (2014)
63. B. Ravel, M. Nervile, J. Synchrotron Radiat. **12**, 537 (2005)
64. Y. Yasuda, A. Tohsan, W. Limphirat, Y. Ikeda, Kobunshi Ronbunsyu, **72**, 16 (2015) [In Japanese]
65. A. Tohsan, R. Usami, R. Kishi, Y. Yasuda, Y. Ikeda, Memoirs of the SR Center. Ritsumeikan Univ. **17**, 135 (2015) [In Japanese]
66. Y. Ikeda, Y. Yasuda, T. Ohashi, H. Yokohama, S. Minoda, H. Kobayashi, T. Honma, Macromolecules **48**, 462 (2015)
67. Y. Ikeda, H. Kobayashi, S. Kohjiya. Kagaku **70**, 19 (2015) [In Japanese]
68. J.J. Low, A.I. Benin, P. Jakubczak, J.F. Abrahamian, S.A. Faheem, R.R. Willis, J. Am. Chem. Soc. **131**, 15834 (2009)
69. G. Parkin, Chem. Rev. **104**, 699 (2004)
70. GaussView 5, Gaussian Inc.
71. I.N. Levine, *Quantum Chemistry*, 7th edn. (Pearson, New Jersey, 2013)
72. F. Jensen, *Introduction to Computational Chemistry*, 3rd edn. (Wiley, West Sussex, 2017)
73. T.S. Kuhn, *The Structure of Scientific Revolutions,* 2nd edn. (University of Chicago, Chicago, 1970). [The first edition of this book was published in 1062, the 3rd one in 1996, and the 50th Anniversary edition was published in 2012]
74. L. Bateman, C.G. Moore, M. Porter, B. Saville, in *The Chemistry and Physics of Rubber-like Substances*, ed. by L. Bateman (Maclaren, London, 1963), Chap. 15
75. A.Y. Coran, ChemTech **13**, 106 (1983)
76. I.R. Gelling, M. Porter, in *Natural Rubber Science and Technology*, ed. by A.D. Roberts (Oxford University Press, Oxford, 1988), Chap. 10

77. R.P. Quirk, Prog. Rubber Plast. Technol. **4**, 31 (1988)
78. F.A. Carey, R.J. Sundberg, *Advanced Organic Chemistry, Part B: Reactions and Synthesis*, 5th edn. (Springer, Berlin, 2007)
79. A.Y. Coran, in *The Science and Technology of Rubber*, 4th edn., ed. by B. Erman, J.E. Mark, C.M. Roland (Academic Press, Waltham, 2013), Chap. 7
80. Y. Ikeda, in *Chemistry, Manufacture and Applications of Natural Rubber*, ed. by S. Kohjiya, Y. Ikeda (Woodhead/Elsevier, Oxford, 2014), Chap. 4
81. Y. Ikeda, A. Kato, S. Kohjiya, S. Takahashi, Y. Nakajima, *Gomukagaku (Rubber Science)* (Asakura Shoten, Tokyo, 2016), pp. 137 [In Japanese]

Chapter 5
Pneumatic Tire Technology

5.1 History of Tire Technology

5.1.1 Invention of Wheel and Development of Rubber Tire

(1) Invention of Wheel

Our hominid ancestors were born in the savannas of Africa and then began their transcontinental journey around the world. The desire for mobility might have been due to their critical survival. It was so critical that the desire for motion might have been embedded in our genetic makeup. The land and water (river, lake, sea, or even ocean) transportations of both people and goods have absolutely been indispensable to human existence. It has been said that the oldest means of land transportation (except walking on foot) was a sled, which is still used on ice and snow today. The sled slides easily owing to the very low friction on ice or even on snow.

Experiencing the lower friction of rolling than that of sliding, people might have recognized the use of a roller (initially wooden logs), which might give rise to the evolution of a wheel to gradually develop the wheel and axle. There have been lots of evidences for the manufacture and use of wheeled vehicles in prehistoric Europe for a period over three and a half millennia. As an initial innovation, it has raised classic questions of monogenesis or polygenesis. For examples, single piece wooden wheel with integral nave of the mid-third millennium BC was excavated, rock carvings of ox-drawn wagon were found dated to the third millennium BC, and Sumerians living at the mouth of the Tigris–Euphrates river system (in present-day Iraq) were supposed to use the wheel around 3500 BC, which is suggested from a sketch depicting a sledge mounted on four solid wheels [], to name a few. The circumferences of these wheels were covered by animal skin fixed to the wheels by copper nails, which we call now a tire. The tires refer to an item that covers the circumference of a wheel and was invented after the wheel. About 2000 years ago, in the days of the Roman Empire, the Celts developed an innovative approach of thermally applying a ring made from iron to the circumference of a wooden wheel. In many countries, the wooden wheels with iron tires have been fitted to large carts and wagons pulled by livestock. Even in Japan, especially in the country side, they had remained to be of use until 50 years ago.

(2) Development of Rubber Tire

Problems of intense vibration and loud noise soon materialized on the unpaved roads, when wagons mounted with wheels having iron tires travelled, although the wooden wheels encircled by iron tires had been in use, especially for carrying a load or freight (not a passenger), for thousands of years. Solid rubber tires (non-pneumatic ones) appeared in 1835. But, the improvement of vibration and noisiness was not much, since the used rubber was raw and was not vulcanized. Also, the wear life of the tires was poor [2]. Furthermore, solid rubber tires had drawbacks in that rubber melted and softened at high temperature in summer, while it was rigid to form cracks at low temperature in winter. In 1839, Charles Goodyear in the USA (who was not a founder of the current Goodyear Tire and Rubber Company) invented a way of overcoming these thermal characteristics. His invention was named vulcanization, in which raw rubber premixed with sulfur and white lead was heated for hours. Following his work, Thomas Hancock in England developed various vulcanization methods. Combined with his experiences on the processing of rubber, he opened the door to the modern rubber industry in England. After the durability and wear were improved by the vulcanization of rubber, solid tires were mainly used for bicycles [3]. Solid rubber tires were still used for military vehicles in World War I beginning in 1914, and it was said that the maximum speed of such military vehicles was 30 km/h. If the vehicles were driven for a long time at their maximum speed, the solid tire was said to emit smoke due to the high temperature of rubber. Solid tires made of vulcanized rubber are still used for small tricycles and baby carriages, since their moving speed is low enough.

5.1.2 Invention of Pneumatic Tire and Its Development

(1) Invention of Pneumatic Tire

The pneumatic tire, which is indispensable to current vehicles, was invented for steam-engine automobiles by Robert W. Thomson in England; he patented the pneumatic tire in 1845. Figure 5.1 shows the configuration of his pneumatic tire, named the aerial wheel, which consists of a tube made of rubber-coated canvas, a wooden rim, a stud, a skin that covers the outside, and air that inflates the tire. The overall construction and the function of each part in this aerial wheel are essentially the same as those of the current tires. It has been said that people were surprised at the good riding comfort, low noise, and low rolling resistance of wagons equipped with the aerial wheels [3–7]. However, difficulty in mounting these tires on rims and the high price of the tires made the use of the tires quite limited, hence not widespread. Furthermore, the Red Flag Act promulgated in 1865 in England

Fig. 5.1 Pneumatic tire invented by Thomson (from Ref. [6])

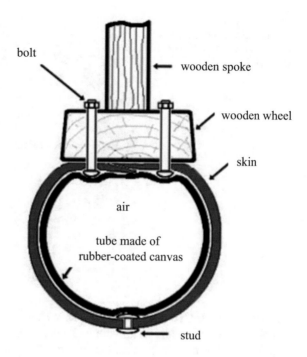

bolt

wooden spoke

wooden wheel

skin

air

tube made of
rubber-coated canvas

stud

requested that steam automobiles must travel at speeds lower than 6.5 km/h and that a guard must walk in front of steam automobiles. This act made the steam automobile unpopular, and Thomson's invention was forgotten.

About 40 years later, in 1888, John B. Dunlop obtained a patent for a pneumatic tire that improved the riding comfort of his son's small tricycle that originally had solid tires. This was a reinvention of Thomson's idea. In 1889, bicycle racers using Dunlop's pneumatic tires overwhelmingly defeated other cyclists racing with solid tires. It became widely known that the rolling resistance of pneumatic tires was so low that a bicycle could be pedaled with less power. A company that produced pneumatic tires, the predecessor of Dunlop Tires, was then founded in Belfast, Ireland [3].

Pneumatic tires were first applied to a motor vehicle by the brothers Édouard and André Michelins (who founded Michelin in 1889) in France. Following the success of pneumatic tires for bicycles, the Michelin brothers drove their self-made car equipped with pneumatic tires in the car race between Paris and Bordeaux (1179 km round trip) in 1895. Their tires punctured so frequently that 24 tubes were used in the race and they were the ninth finisher of the nine cars that completed the race. Even though they ranked at the last in the race, they declared that all motor cars would be equipped with pneumatic tires within 10 years and began selling pneumatic tires for vehicles [3, 4].

For the widespread use of pneumatic tires, they had technically to improve the assembly much, i.e., mounting and removal of pneumatic tires to and from wheels

Fig. 5.2 Invention for assembling a pneumatic tire to a wheel invented by Welch (from Ref. [6])

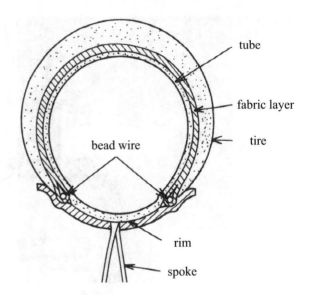

needed to be easier. To this end, Charles K. Welch invented the tire bead assembly. It was made of wires and a rim with a circular cross section in 1890 as shown in Fig. 5.2. His invention was called the "wired on" or "straight sided" assembly. The same year, William K. Bartlett invented the "clincher" assembly in which stiff rubber hung on the flange instead of wires. The current bead construction of tires was completed by further developing the inventions of Welch and Bartlett [3, 8].

(2) **Progress in Pneumatic Tire Technology**

The performance of pneumatic tires has been improved by many inventions. Historically, great inventions for tires following the aforementioned ones are the following:

(i)   Change from woven canvas to a cord fabric where warps of strong tire cords are weaved with wefts of relatively weak cord (in 1892)
(ii) Reinforcement of tire rubber by carbon black (in about 1904)
(iii) Application of synthetic fiber to the tire cord (e.g., Nylon in 1942)
(iv) Advent of the steel belted radial tire (in 1946) [3, 8].

These technologies are briefly explained:

(i) Canvas for clothing was used in tires as skeletal material in the early days. Because warps and wefts crossed each other in the canvas, both underwent rubbing due to deformation, whenever a tire rolled and they rapidly wore out. In 1892, John F. Palmer in the USA applied a patent of a cord fabric in which cords were oriented in parallel within a layer, and thin rubber was inserted between layers corresponding to the warp and weft and the layers were pasted together. The new construction managed to solve the rubbing problem between the warp and weft, and it spreads over Europe after World War I. Thus, the tire life improved from at most 5000 km to three to five times of this distance [3].

(ii)   In about 1904, Sidney C. Mote in England discovered that the physical properties of rubber were outstandingly improved by adding carbon black [9]. This discovery was first applied to tires in about 1912; the durability of a tire was improved 10 times by adding carbon black [2, 3]

(iii) Cotton cords were used for tires in the early days, but tire manufacturers were urged to improve tire performance as the driving speeds increased with better road maintenance. Hence, rayon and nylon were first applied to tires in 1938 and 1942, respectively, in the USA. As a result, the weight of tires was reduced to result in the improved high-speed durability. Aramid fiber (Kevlar) was then applied to tires in the 1970s [8]. These synthetic fibers are still used in various tires, rubber belts, and some other rubber goods for different purposes.

(iv)   The fourth historic invention, the steel belted radial tire, is estimated to be the most important technological innovation in the history of pneumatic tires and is described here in detail. At first, two types of rubber tire are introduced here. A bias tire and a radial tire are compared in Fig. 5.3. Carcass of the bias tire consists of cord fabrics at opposing angles. Because the direction of cords is oblique, this type of tire is called the bias tire. There used are two to dozens of fabrics, depending on the inflation pressure. Meanwhile, the carcass structure of the radial tire consists of cords whose direction is approximately radial when seen from the lateral side of the tire, which is the reason why it is called a radial tire. If we inflated a tire having only radial plies, the plies would split each other owing to the stretch in the circumferential direction. Hence, a steel belt that functions as a hoop is combined with radial plies in the radial tire.

The first patent relevant to the steel belted radial tire was obtained in 1913 by two Englishmen, Christian H. Gray and Thomas Sloper [4]. Unfortunately, their

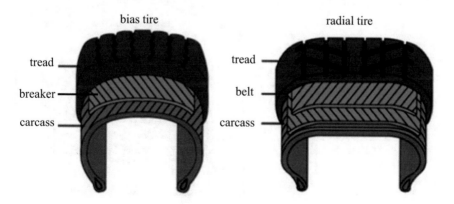

Fig. 5.3 Comparison of a bias tire and radial tire (from Ref. [1])

idea was not immediately implemented owing to the outbreak of World War I. In 1939, Marius Mignol from Michelin studied the contribution of each part of a bias tire to the rolling resistance of the tire. He intentionally made a radial tire without a belt and found that the contribution of the side wall to the rolling resistance could be negligible for a radial tire. Because rolling losses both in the tread and side wall were very slight in this tire, Pierre M. Bourdon from Michelin began research on the radial tire. He developed a steel belted radial tire for locomotives and validated that the radial tire was superior to the bias tire in many ways [4]. After a patent application in France in 1946, a small-scale production of radial tires started in 1948. The manufactured steel belted radial tire was named X and was selected as the standard tire for Lancia, and then for Alfa Romeo. It was also a recommended tire for Ferrari [4, 5]. This tire was made of three belts as shown in Fig. 5.4, and there observed drastic improvements in rolling resistance, durability, and maneuverability. The X tire had drawbacks in terms of poor riding comfort and spinning during severe cornering. However, Michelin managed successfully to solve these problems of the X tire by adopting a new belt structure, in which the number of belts was reduced from three to two and a stiff rubber layer was sandwiched between the two belts to improve the stiffness of the tread. The new tire, named ZX, went on sale in 1967 and enjoyed wide popularity among European customers. The structures of present tires are fundamentally based on the structure of the ZX tire [2, 4]. Afterward, the tire industry has made a series of smaller inventions to improve the basic structure of the ZX tire.

Among the major inventions after the radial tire were the invention of the tubeless tire by Frank Herzegh at B.F. Goodrich in 1947 and the concept of all-season tires by Goodyear in the 1970s [8].

Fig. 5.4 Three-belt structure of the radial tire X produced by Michelin (from Ref. [2])

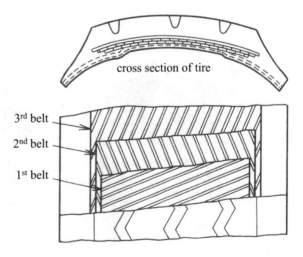

cross section of tire

3rd belt

2nd belt

1st belt

5.1.3 Technological Innovation of Pneumatic Tire

(1) Progress in the 1990s

Three technological innovations in the 1990s were mentioned here. That is, (i) the application of silica filler to tires, (ii) the development of run-flat tires, and (iii) fully automatic tire building systems.

(i) It was recognized in the 1970s that if silica, instead of carbon black, was well mixed in rubber, the energy loss of the rubber decreased. But, also known were the deterioration of durability and lower wear performance. Lots of chemical and tire companies carefully studied the existing research reports and activity on silica, and they failed to put the research and developmental trials into practical applications of silica to tires. But, Michelin made it again! In 1992, it launched the "green tire," where carbon black was partially substituted by silica in the tread rubber, to achieve lower fuel consumption. The green tire decreased rolling resistance by 20% without a change in the friction coefficient for a wet road. Counteracting the global warming problem by the exhaust gas from automobiles and the higher price of gasoline, silica technology seemed to give a kind of compensation to the tire industry, in terms of solving the incompatibility between rolling resistance and braking distance on a wet road. The president of Michelin, however, mentioned at the time that it would take more than 20 years to develop the silica technology, to the dismay of many tire companies. The wear of silica-loaded rubber is still worse than that of carbon-black-loaded rubber, which still remains to be a big problem for silica technology.

(ii) Studies on a run-flat tire system began in the early 1900s. Many ideas were proposed; e.g., multiple tubes inside a tire, foam inside a tire, the automatic injection of a sealant to repair a puncture, elastic structure mounted inside a wheel to support a load in the event of a puncture, and a reinforced side wall, to name several. In the 1970s, some of these ideas were applied to the tires of military vehicles and then adopted as optional tires for passenger cars. However, run-flat tires were not widely used owing to the irregular wear of tires, which might be caused by the poor shape of the tire footprint of the run-flat system [5]. Following improvements, run-flat tire systems are gradually being accepted by car manufacturers.

Current run-flat tire systems can be classified into the three categories shown in Fig. 5.5. The International Organization for Standardization (ISO) defines run-flat tires as tires that can run 80 km at a speed of 80 km/h under a zero inflation pressure state. The most popular run-flat tire system so far is a self-support run-flat tire having a side wall that is reinforced by stiff rubber. Inner-ring systems are classified into two: the system with a standard rim and that with a special rim. Michelin promoted an example of the latter run-flat

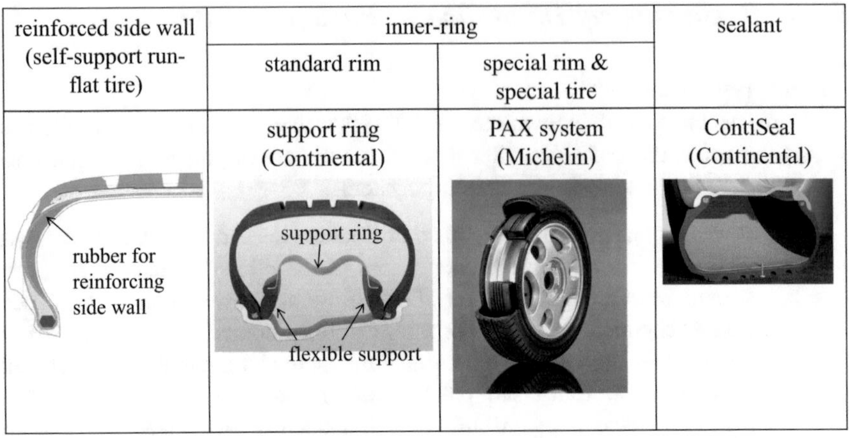

reinforced side wall (self-support run-flat tire)	inner-ring		sealant
	standard rim	special rim & special tire	
rubber for reinforcing side wall	support ring (Continental)	PAX system (Michelin)	ContiSeal (Continental)

Fig. 5.5 Three categories of run-flat tire system (from Ref. [11])

system named PAX, in particular, but it was not successful owing to a difficulty in mounting tires on a special rim and other problems. Run-flat tires with sealant first went on sale in the 1970s but were not successful owing to instability of the sealant. In recent years, Continental has begun selling run-flat tires with sealant again. In any case, it is notable that a vehicle requires a tire pressure monitoring system (TPMS) which informs a driver of a tire puncture, if a vehicle is equipped with any type of run-flat system. Although the tire industry is recommending installment of the run-flat system, the system is only adopted by a few car companies, like BMW. Owing to stiff riding comfort, high rolling resistance, high price, and requirement for a TPMS, run-flat tire systems are not widely used yet.

(iii) In the 1990s, Michelin developed a fully automatic tire building system named C3M, which was different from previous systems, and the other tire companies followed Michelin in developing similar systems. Although tire uniformity is improved by adopting the fully automatic tire building system, the system is not so popular as expected owing to many problems. For example, there is a waste of rubber material when the type of rubber material is changed for the sake of automation, the system is not as flexible as expected, and it is difficult to tune tire performances appropriately because the fully automatic tire building system has limited engineers' imaginative and original works in terms of designing tires.

(2) **Effect of the Innovations on Tire Industry**

The aforementioned inventions and innovations on tire technology can be classified into three areas of technological change as shown in Table 5.1: technological changes relating to the material, construction, and tire building [10, 11]. The technological change that had the greatest effect on the tire industry may be the change from a bias tire to a steel belted radial tire. This is supported by the fact that four of the five large

Table 5.1 Effects of technological innovations on the tire industry

Year	Technological contents	Areas of technological change			Effect on tire industry
		Material	Construction	Tire building	
1892	From canvas to cord fabric	C	A	B	C
1904	Reinforcement of tire rubber by carbon black	A	C	B	C
1842	Usage of nylon	A	C	A	C
1946	Steel belted radial tire	A	A	A	A
1990s	Silica-loaded tire	A	C	B	C
	Run-flat tire	B	A	C	C
	Fully automatic tire building system	B	B	A	B

From Ref. [10, 11]
Degree of influence: *A* large; *B* medium; *C* small

tire companies in the USA could not adapt to the change and were overwhelmed by other companies [12]. The development of the steel belted radial tire is usually discussed in terms of the change in the carcass structure from bias to radial, but we must understand that there were three changes.

(i) Change in material: A steel belt as a new material was applied to a tire and the adhesion between steel cord and rubber had to be studied.

(ii) Change in construction: The carcass structure was changed from bias to radial.

(iii) Change in tire building: A machine for building the radial carcass and belt had to be newly installed in a factory.

Some tire companies in the USA were too large to adapt to these technological changes because the three changes occurred at the same time in the development of a steel belted radial tire. Furthermore, it was an unfortunate coincidence for the four overwhelmed tire companies that the recession caused by the oil crisis in the 1980s occurred when they needed to invest in a steel belted radial tire.

Among technologies developed in the 1990s listed in Table 5.1, the fully automatic tire building system may be the second most influential technology for the tire industry. If one could solve the problems with the system, the ordinary tire building system would become obsolete. In such a scenario, small tire companies would gain a competitive positioning, because relatively they do not possess many ordinary tire building systems which would become worthless after introducing the new tire building system. The manufacturing process of an ordinary tire building system is not explained here. Refer to the literature [13] in which the process is simply explained with figures.

5.2 Function of Pneumatic Tire

5.2.1 Four Functions of Tire

Tires have to show four basic functions:

(i)   Tires support the vehicle's load (i.e., they have a load support function).
(ii) Tires transmit braking and acceleration forces to the road (i.e., they have a traction and braking function).
(iii) Tires maintain and change vehicle direction (i.e., they have a maneuverability function).
(iv) Tires absorb shocks from road surface (i.e., they have a riding comfort function).

The whole description of this chapter suggests that pneumatic rubber tires have been the best choice in executing these unique functions so far, which is briefly explained below.

While a solid tire supports a vehicle's load through the deformation of rubber, a pneumatic tire supports the load by the pressure of compressed air inside the tire as well as by the deformation of rubber. As shown in Fig. 5.6, the downward component of carcass tension is small near the contact patch owing to deformation of the tire, while the upward component of carcass tension does not change near the upper region of the loaded tire. Because a pneumatic tire largely deforms only near the contact patch, the resultant force T of the vertical component of the carcass tension is generated and partially supports the load W. Furthermore, shocks are

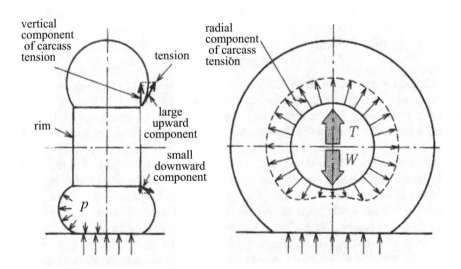

Fig. 5.6 Mechanism of how a pneumatic tire supports a load (from Ref. [2])

absorbed because of the low spring stiffness of a pneumatic tire resulting from large deformation near the contact patch.

When a driver turns a wheel to the right, the center plane of the front wheel also turns to the right relative to the moving direction of the vehicle as shown in Fig. 5.7. A pneumatic tire largely deforms in the lateral direction near the contact patch, and the cornering force is generated as the reaction to this lateral deformation. Because a pneumatic tire can deform largely in the lateral direction, the cornering force is relatively linear up to a large slip angle. Meanwhile, the lateral deformation of a solid tire or steel wheel is extremely small, and the cornering force is thus easily saturated at a small slip angle. Hence, a vehicle with solid tires is difficult to drive.

The maneuverability of a steel belted radial tire is better than that of a bias tire. This is because the lateral force of a steel belted radial tire is stronger than that of a bias tire, owing to the stiff steel belt. In braking or acceleration, a tire deforms in the direction of movement, which is different from the direction of deformation during cornering in Fig. 5.7, but the tire performance during braking or acceleration can be explained in the same way as the tire performance during cornering. Hence, the performance of a steel belted radial tire in braking or acceleration is also better than that of a bias tire.

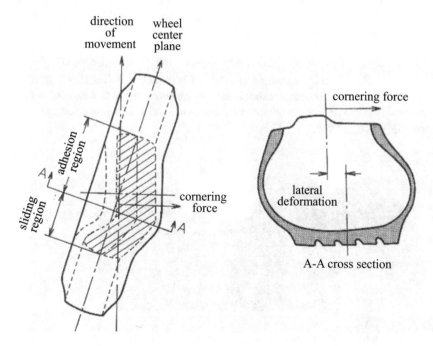

Fig. 5.7 Deformation of a pneumatic tire in cornering (from Ref. [2])

Car manufacturers are developing vehicles with improved fuel efficiency and lower emission of carbon dioxide, in response to environmental issues such as the global warming, a steep rise in fuel prices, and the drain on resources. Hence, the tire industry has accelerated their studies on the lower rolling resistance and wear of tires, which are related to global warming and use of the limited resources, in addition to decrease of tire noise, which is related to the quality of life. If we refer to these issues as the important environmental performance, we can say that tire research has been shifting from the study of basic functions to that of the environmental performance. This change may thus be a paradigm shift in the definition of the basic functions of a tire, which can continue to the end of this century.

5.2.2 Elements in Designing Pneumatic Tire

A tire, which is a part of a vehicle, is expected to satisfy multiple performances simultaneously when moving. This requirement has to be fulfilled only by its composite structure, i.e., it is assembled from multiple composite materials. Figure 5.8 shows that composite materials are used in the tread of tire: The tread rubber is a composite of cross-linked rubber mixed with a reinforcing nanofiller, such as carbon black, and the rubber and steel cord or organic fiber are in special combination to form tread, which comes in contact with the road. Also, a composite material made of rubber and organic fiber is used in the side walls of passenger-vehicle tires, while rubber and steel cord are used in the side walls of truck/bus tires. Note that the used rubber itself is microscopically a nanofiller-reinforced rubbery composite made of raw rubber, sulfur curing system (sulfur, accelerator, activator), antioxidant, antiozonant, carbon black or silica nanoparticles or sometimes both for reinforcement, and a few other reagents for improving processability.

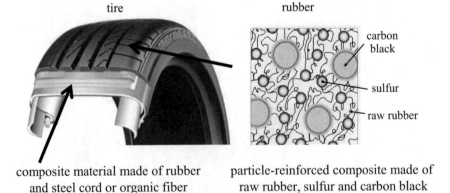

Fig. 5.8 Tire and composite materials (Prepared by Y. Nakajima)

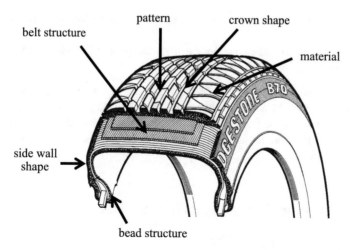

Fig. 5.9 Design elements of a tire (Prepared by Y. Nakajima)

There are four elements in designing a tire: the shape (relating mainly to the side wall and crown), construction (relating to the belt and bead), tread pattern, and the materials used (Fig. 5.9). The difficulty or the essence in tire design is to satisfy the aforementioned four basic functions of tire and the environmental performance, while using only four design elements. Table 5.2 gives the contributions of design elements to tire performance.

A typical process of tire design is including the steps as follows:

(i)   Tread patterns of good aptitude are designed beforehand because it takes a long time to develop a pattern.
(ii) Shapes of the side wall and crown are designed to satisfy the target performance, and tire molds with the pattern are made.
(iii) Performance is finely tuned to satisfy the target performance by changing the construction and material.

Table 5.2 Contribution of design elements to tire performances

Design element	Rolling resistance	Brake	Wear	Riding comfort	Maneuverability	Noise	Number of design parameters
Shape	B	C	B	B	B	C	About 10
Pattern	C	A	A	C	A	A	A few hundreds
Construction	B	B	A	B	A	C	About 50
Material	A	A	A	B	A	B	A few hundreds

Prepared by Y. Nakajima
Degree of contribution: *A* large; *B* medium; *C* small

(iv) Performance is evaluated in not only indoor and outdoor tests but also in the market.

(v)   If the target performance is not satisfied, the pattern is revised and the process is repeated from the second step.

The duration for tire development may be a few years in the case of passenger-vehicle tires and probably more than 10 years in the case of truck/bus tires, where the target performance level is higher. Past and current technologies for the shape, construction, pattern, and material of the tire design are explained in the following section.

5.3   Engineering Design of Pneumatic Tire

5.3.1   Shape Design of Tire

(1) Progress of Tire Shape Design

Tire shape of the side wall has been theoretically studied by many researchers in the development of tire technology. Because the difficulty of design is proportional to the number of parameters in each design element, tire shape of the side wall is the easiest theme among tire design elements as suggested in Table 5.2. It has been said that the first study on the shape of the tire side wall was carried out by Schippel [14]. Expressing the tire shape using an ellipse, he derived an equation of the tire shape in accordance with an equilibrium between the inflation pressure and the carcass tension. In 1928, Day and Purdy [15] of Goodyear started theorizing the inflated shape of a bias tire but they did not disclose their theoretical details. Hence, it is not exactly known when the theory of the shape of a bias tire was completed and when it was applied to a tire. A few other tire companies also performed similar studies at that time. The theorized shape is called the natural equilibrium shape, and it is still used as a reference shape when a tire shape is designed. Purdy [16] later wrote an excellent book on tire shape, which is still of much use as an initial reference.

The tire shape of a radial tire was first theoretically studied by Böhm [17]. Based on the fact that both the belt and carcass support air pressure in a radial tire, the equation for a radial tire was derived by assuming that the allotment of air pressure to the belt is parabolic in the width direction. Yamagishi et al. [18] proposed a non-equilibrium shape that allows the control of the tension distribution of the belt and the bead areas. They validated that the maneuverability, rolling resistance, and wear of passenger-vehicle tires were improved by enhancing both the belt and the bead tensions through the non-equilibrium shape. A similar idea was applied to truck/bus tires by Ogawa et al. [19]. However, they failed to derive an equation or some other procedures to define the non-equilibrium shape.

Nakajima et al. [20] solved this problem by developing a theory that derives the optimized tire shape using the optimization technique and the finite element method

(FEM). In this theory, the tire shape was obtained from the optimization result. Furthermore, any tire performance would be improved by this theory if the objective function for tire performance could be defined. This theory could be an ultimate theory on the shape of the side wall, as long as the tire construction was as shown in Fig. 5.9. Also, Nakajima et al. [21] applied a similar idea of optimization to the tire crown shape, which would be the ultimate theory on the shape of the tire crown. The optimization procedure has become the standard approach for tire design in the industry, and in terms of application of optimization technology to product design, the tire industry may thus enjoy the most efficiency.

(2) Optimized Design of Tire Shape

Figure 5.10 compares the optimized shape and a control shape of a side wall of a passenger-vehicle tire. The objective function in optimization is the maneuverability of the tire, which is improved by increasing the belt and the bead tensions of an inflated tire. The constraints of the optimization are to keep the tire size within the allowance of regulations and not to deteriorate other performances. The optimized tire shape is an unprecedented wavy shape that no tire designers have ever imagined! This surprise is due to the imprinted tire shape on the designer's mind that it should protrude outwardly as shown in Fig. 5.10 (see the control shape). Maneuverability performance was evaluated in an indoor test and subjective vehicle test in a proving ground, and it was validated that improved maneuverability, rolling resistance, and the other performances were achieved by the tires of the optimized shape.

A similar method was applied to the optimization of the crown shape, where the objective function was assumed to make the contact pressure uniform. The maneuverability and wear, especially irregular one, were much improved by the optimized crown shape [21].

An unprecedented design might be proposed through hundreds of computer simulations during optimization, but it does not mean that the past experiences and

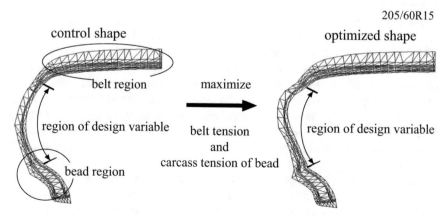

Fig. 5.10 Comparison of the control shape and optimized shape of the side wall (from Ref. [20])

experiments are of no use. An optimized design system might have made the tire designers free from employing a trial-and-error approach and have provided them with a scientific insight for improvement. When the current design process is systematically incorporated into the constructed optimized computer system, majority of accumulated tacit knowledge of tire designs would be transformed into as explicit knowledge leading to the further integration of current knowledge.

5.3.2 Structural Design of Tire

The structural design of a tire is a process in which target tire performance is satisfied by controlling the configuration (belt width, belt angle, rubber thickness, etc.), stiffness, and viscoelastic properties of fiber-reinforced rubber (FRR) and rubber materials. FRR is a rubber material reinforced by fibers or steel cords and has been widely used in flexible products such as tires, belts, and hoses even before the term composite material was first used. Comparing FRR with fiber-reinforced plastic (FRP), which is widely used as a composite material, the ratio of elastic moduli between fiber or steel cord and rubber as a matrix ranges from 100 to more than a few thousands in FRR, while it is about 10 in FRP owing to the much higher modulus of plastic as a matrix. The difference in the ratio of elastic moduli causes that the anisotropic behavior of FRR is different from that of FRP. For example, in the laminated bias unidirectional cord reinforced composite used for passenger tires, there is the particular bias angle in FRR where the shear deformation is disappeared in extensional load, while it is not the case in FRP [22]. Furthermore, the other characteristics of FRR are different from those of FRP. For example, the matrix of FRR is a kind of isotropic composite in which particulate nanofiller is mixed for reinforcement, while that of FRP is usually plastic materials only with additives of low molecular mass. Also, fiber or steel cord in FRR is twisted, while short or long fiber without twist is used in FRP.Please check the clarity of the phrase 'ranges from 10 to in FRR' in the sentence 'Comparing FRR with...'. I added the number to correct the sentence.

Structural designs of a tire are classified into two; the belt construction design and the bead construction design. The belt of a tire is made of multiple laminated FRRs reinforced by steel cords at angles to the circumferential direction. Many types of laminated structure (including topological structures) and various materials of cord and rubber are used to achieve the target performances. In passenger-vehicle radial tires, the belt is made of a two-layer composite reinforced by steel cords and the carcass is made of one- or two-layer composite reinforced by organic fibers. Meanwhile, in truck/bus radial tires, the belt is made of three- or four-layer composite reinforced by steel cords. In some earth-mover tires, the belt is even made of a seven-layer composite.

Nakajima et al. [23, 24] proposed a new design procedure for the belt and bead constructions in which the optimization technique was combined with the FEM. A new belt structure, which is topologically different from the control belt structure,

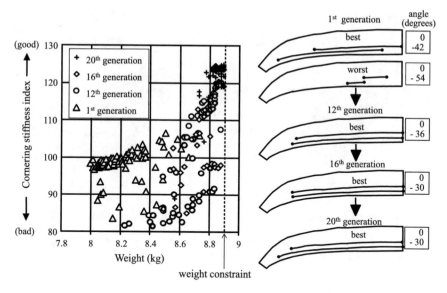

Fig. 5.11 History of evolution of belt construction in GA (from Ref. [24])

could be generated using this technology. This optimized design procedure may indicate the direction of ultimate structural design.

Figure 5.11 shows the optimized results to maximize the tire's cornering stiffness in the constraint of tire weight by the GA (genetic algorithm). The GA is search algorithm based on the mechanics of natural selection and natural genetics. In the GA, the evolutionary process of living things is grasped, the process of survival of the fittest. The right figures of Fig. 5.11 show the best and the worst construction and the corresponding belt angles in the population at each generation. Because the initial constructions (first generation) are generated by random numbers, various belt structures are generated. The left figure shows the cornering stiffness index and the weight of tires of the whole population. The GA gradually improves the belt construction until the optimized construction is generated at the twentieth generation. Though a separated belt construction is generated in the early stage, the optimized structure is topologically the same as the commonly used structure. The tire with optimized structure was built, and the cornering stiffness was measured using a drum tester. The optimized structure was validated to increase the cornering stiffness more than 15 percent as compared with the commonly used structure in an indoor drum test. Figure 5.12 is the other application of the optimization for bead structure in which the shape of bead filler is optimized to improve the durability of tires with low turnup construction. The principal strain at the end of carcass is decreased by 18% in a loaded tire by the optimized bead structure, and the durability of the tire was drastically improved in the drum test.

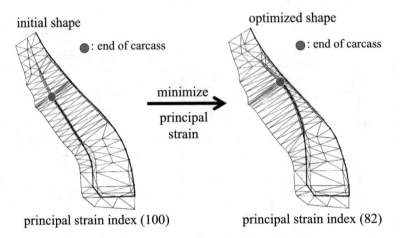

initial shape

⬤ : end of carcass

minimize

principal

strain

optimized shape

⬤ : end of carcass

principal strain index (100)   principal strain index (82)

Fig. 5.12 Optimized shape of bead filler to improve durability of tires with low turnup construction (from Ref. [23])

5.3.3 Design of Tread Pattern

The tread pattern of a tire consists of wide grooves and narrow grooves called sipes, and it importantly affects various tire performance requirements as shown in Table 5.2. The tread pattern is usually designed according to the following four steps:

(i)   A pattern designer creates many original pattern designs according to a concept by considering the appearance, pattern stiffness, and length of sipes.
(ii) A few patterns of good aptitude are selected from these candidates through evaluation of noise, wear, maneuverability, etc., by using computer aided engineering (CAE) tools for a tire pattern such as analytical tools and numerical simulations.
(iii) Trial tires with selected patterns are prepared, and their performance is evaluated by indoor and outdoor tests.
(iv) If the target performance is not satisfied, the procedure returns to step (ii) and the patterns are revised. The process is repeated until the target performance is satisfied.

The role of the tire pattern is to improve the grip of road surface, particularly under wet, snowy, and icy surface conditions. However, some carved patterns of the tread rubber may result in noise and irregular wear. Many CAE tools have been developed to predict the performances of different tire patterns, and the most popular method is to predict tire noise according to theory on pitch noise developed by Nakajima [25]. Furthermore, a CAE tool using the FEM to predict the interaction between the tire and road is often applied in the development of tire patterns. For example, Nakajima et al. [26, 27] first developed a practical FEM simulation for hydroplaning, which afforded a few practical model patterns, and the simulation showed useful predictability in tire pattern designing, that is, the predicted water flow in the contact area agreed well with measurements, and the raking of the critical speed of hydroplaning could be predicted well for tires with different patterns.

The hydroplaning simulation using the FEM was further employed to develop the pattern of a Formula One (F1) tire to be used in wet conditions. Due to the instrumental difficulty and the limited space, the evaluation of hydroplaning of an F1 tire was done not by an indoor test, but the tire evaluation was developed by repeating outdoor vehicle tests. A CAE tool for hydroplaning allowed tire engineers to see water flow in the contact patch of the F1 tire. Because water flow indicated where water tends to flow out of the contact patch, water might be effectively drained if the pattern was grooved along the predicted water flow. The predicted water flow of a front tire with blank tread was asymmetric owing to the camber angle, while that of a rear tire was symmetric as shown in Fig. 5.13. The newly developed pattern geometry is shown in Fig. 5.14 where the grooves for a front tire and a rear tire gradually change from the circumferential direction in the center area to the lateral direction in shoulder area. The groove configuration became asymmetric in a front tire, and the centerline of the contact area was not located at the center of the tire geometry owing to the camber angle. In 2000, Michael Schumacher drove a Ferrari F1 car with the new tire developed by the CAE tool in

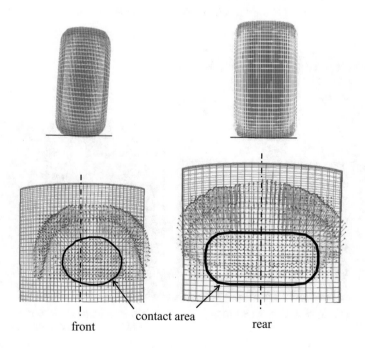

front   contact area   rear

Fig. 5.13 Water flow of tire with blank tread (from Ref. [27])

Fig. 5.14 New pattern
geometry for F1 tires (from
Ref. [27])

the rain at Silverstone in England, and comprehensively defeated competitors with
other patterns. The race demonstrated that the CAE tool for hydroplaning was able
to develop a practically excellent pattern.

By further progressing of this technology, a simulation for tire traction on snow
was conducted [28]. Figure 5.15 shows the shear force in the grooves between the
tire and snow, which could not be measured in an experiment, could be well
visualized by the simulation technique. Because this technology has been of use to

(a) Predicted snow surface   (b) Predicted shear stress
                                          distribution at contact patch

Fig. 5.15 Visualization of shear force in grooves between tire and snow (from Ref. [28])

make the development of a pattern for studless tires possible even in a season when an outdoor test on snow is difficult, it has become an indispensable technique in pattern designing.

5.4 Material Design of Pneumatic Tire

5.4.1 Materials for Tire

From the view point of tire performance, the material is the largest contributor among the design elements as shown in Table 5.2. An automotive tire is made of more than 100 components, such as raw rubber, tire cord, carbon black, bead wire with lots of compounding agents mainly mixed in raw rubber. Half of the raw materials are chemical products made from petroleum (naphtha), and the production of tires thus depends highly on petroleum. In terms of raw materials by weight, the composition of tire manufactured in Japan in 2014 was: approximately half was rubber (natural rubber 29% and synthetic rubber 22%), reinforcing agent such as carbon black and silica 25%, and tire cord such as steel cord and organic fiber cord 14%, as shown in Fig. 5.16 [29].

The required functions of a tire are satisfied using different rubber materials suited to the different tire parts (see Fig. 5.9). Therefore, as shown in Table 5.3, many types of rubber are used in a tire, e.g., styrene-butadiene rubber (SBR), butadiene rubber (BR), natural rubber (NR), isobutylene-isoprene rubber (IIR), and isoprene rubber (IR) [30]. It is notable here that many rubbery products are man-ufactured from a rubber blend, i.e., a rubber mixture of two or sometimes even three kinds of rubber. SBR, for example, is used for the tread of a passenger-vehicle tire as a main component with BR as the second component, while NR is used as a most major component (sometimes 100% NR) for the tread of a truck/bus and aircraft tires because the stress–strain of the tread of a truck/bus tire is greater than that for a passenger-vehicle tire. IIR is uniquely used for the inner liner, which is the innermost part of a tire. Air permeability of IIR is the lowest among all the

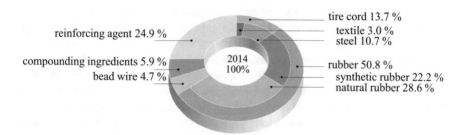

Fig. 5.16 Weight composition of raw materials in tire (from Ref. [29])

Table 5.3 Types of rubber used for different parts of a tire

Part of tire	Tire performance	Passenger-vehicle tire	Truck/bus tire
Tread	Wear, rolling resistance, brake, maneuverability	SBR (styrene-butadiene rubber) BR (butadiene rubber)	NR (natural rubber) SBR, BR
Belt	Durability, stiffness	NR, IR (isoprene rubber)	
Side wall	Flexibility, ozone-resistant	NR, BR	
Inner liner	Air permeability	IIR (isobutylene-isoprene rubber)	

From Ref. [30]

commercially available rubbers; thus, IIR has been and is the best choice for maintaining the air pressure of tire.

Among all the parts of a tire, the material design for the tread rubber is the most important for determining the driving performance of the automobile. Because tread rubber is in contact with the road, it affects many aspects of tire performance, such as rolling resistance, wear, braking distance, wet-skid resistance, and maneuverability. Tread rubber is designed according to the empirical relationship between tire performance and viscoelasticity. Since many tire functions are related to its friction on the road surface, the loss tangent expressed by tan δ is of at most importance. Look at Fig. 5.17 carefully. ICE, WET, DRY, and RR in the figure indicate the braking performances on icy, wet, and dry roads and rolling resistance, respectively [31]. Each performance is mainly controlled by the value of tan δ in the corresponding temperature region.

For example, because the rolling resistance (RR) of a tire is related to the deformation of tread rubber under rotation on the road, the frequency of the external force acting on the tread rubber ranges from 10 to 10^2 Hz depending on the rotational speed of the tire. Meanwhile, the input frequency of the external force acting on the tread rubber for braking performance measured according to the stopping distance of a vehicle ranges from 10^4 to 10^6 Hz. This frequency is much higher than the input frequency for rolling resistance because rubber is microscopically deformed as it follows small road asperities. Comparison of two tan δ curves, "previous polymer" and "chain-end modified polymer," is given in Sect. 5.4.2.

The viscoelasticity of rubber is usually evaluated at an input frequency of 10 Hz. Because the temperature of tire tread and the input frequency for a rolling tire range from 50 to 60 °C and from 10 to 10^2 Hz, respectively, the rolling resistance can be evaluated by tan δ measured at temperatures ranging from 50 to 60 °C and a frequency of 10 Hz. Tire temperature is related to rolling resistance, but road temperature is related to braking performance. Hence, the temperature and input frequency for a sliding tire on a wet road are 30 °C and from 10^4 to 10^6 Hz. Using the time temperature reduction law (i.e., the Williams–Landel–Ferry equation) popularly used for viscoelastic material, a temperature of 30 °C and input frequency ranging from 10^4 to 10^6 Hz can be converted to a temperature of 0 °C and a

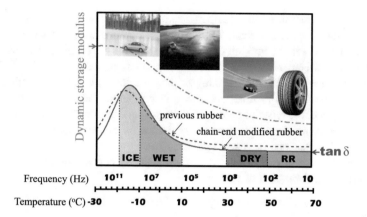

Fig. 5.17 Tire performances and viscoelastic properties of rubber (from Ref. [31])

frequency of 10 Hz. Hence, the braking performance on a wet road can be evaluated by tan δ measured at a temperature of 0 °C and frequency of 10 Hz. Because road temperatures differ among wet, icy, and dry roads, the temperature of tan δ for the braking performance on icy and dry roads is different from that for the braking performance on a wet road. Braking performance is improved by increasing tan δ, while rolling resistance is improved by decreasing tan δ. Figure 5.17 shows that material design to improve braking performance on a wet road and rolling resistance at the same time is to increase tan δ about 0 °C while decreasing tan δ from 50 to 60 °C.

Takino et al. [32–34] showed that the friction coefficient for a wet road measured using a British pendulum tester could be expressed by tan δ measured at a temperature of 7 °C and frequency of 10 Hz and that the cohesion or adhesive force between the rubber and road could be estimated from the wear of rubber in the British pendulum test. Futamura [35] proposed an equation for the rolling resistance and braking distance in terms of not only tan δ but also the modulus of the tread rubber:

$$\text{Rolling resistance or braking distance} = D \cdot E'' / (E^*)^n + F \qquad (5.1)$$

where D, F, and n are constants determined by curve fitting of Eq. (5.1) to the measured rolling resistances and braking distances of several tread compounds. n is called the deformation index, and E'' and E^* are the loss modulus and complex modulus, respectively. The physical meaning of Eq. 5.1 is the energy loss related to various types of deformation, which can be expressed by changing the value of n. For example, $n = 0$, $n = 1$, and $n = 2$ correspond to the deformation with constant strain, constant energy, and constant stress, respectively. The value of n used to predict the rolling resistance and braking distance on dry and wet roads can be determined by the correlation analysis. Five types of rubber were prepared and their properties were measured at a temperature of 50 °C for the rolling resistance and braking distance on a dry road, while rubber properties were measured at a temperature of 0 °C for the braking distance on a wet road. The highest coefficient of correlation between measurements and the prediction made using Eq. (5.1) was obtained for $n = 0.5–1.1$ in predicting rolling resistance, $n = 1.8$ in predicting the braking distance on a dry road, and $n = 0$ in predicting the braking distance on a wet road. Hence, the types of deformation correspond to constant energy for rolling resistance, constant stress for the braking distance on a dry road, and constant strain for the braking distance on a wet road [35, 36].

The energy loss of rubber related to viscoelasticity consists of three microscopic factors in the deformation of rubber:

(i) loss due to the change in arrangement of filler particles loaded in rubber (Payne effect);
(ii) loss due to thermal vibration at free terminals of polymer chains;
(iii) frictional loss between segments in polymer chains.

The contributions of the first and second factors to tan δ are large at temperatures ranging from 50 to 60 °C, which is the temperature region corresponding to rolling resistance. Because the reduction of rolling resistance may be the most important issue for the tire industry, technologies of material design related to the first and second factors are now discussed in detail.

5.4.2 Viscoelastic Behaviors by Rubber Blending and by the End—Groups of Rubber Chains

Rubber is a polymer of very low glass transition temperature T_g, and its tan δ decreases at the temperature range between 50 and 60 °C. For example, T_g of SBR most popularly used in passenger-vehicle tires can be reduced by decreasing the content of styrene or 1–2 butadiene bonds (i.e., vinyl groups). However, the viscoelastic curve in Fig. 5.17 is simply translated laterally without changing the shape by decreasing T_g of the rubber. Therefore, if the value of tan δ decreases at temperatures ranging from 50 to 60 °C, which are related to rolling resistance, the value of tan δ also decreases at a temperature of 0 °C, which is related to braking performance on a wet road. Hence, to satisfy both lower tan δ in the temperature range from 50 to 60 °C and higher tan δ at a temperature of 0 °C, two types of rubber must be blended. Figure 5.18 shows the effects of rubber blends with different T_g on tan δ and the effect of compatibility on tan δ [37]. When rubbers with good compatibility and different T_g are blended, the property of tan δ of the blended rubber becomes the average of the values of tan δ for the two rubbers. Meanwhile, when rubbers with poor compatibility and with different T_g are blended, tan δ of the blended rubber is similar to tan δ of the rubber with low T_g at low temperature and

Fig. 5.18 Effects of a rubber blend with different values of T_g on tan δ (from Ref. [37])

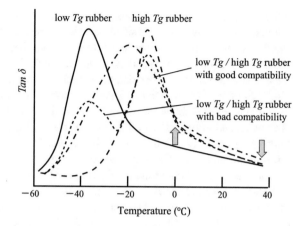

tan δ of the rubber with high T_g at high temperature. Values of tan δ at two temperatures can be controlled by blending two rubbers with poor compatibility.

Chain-end modified polymer and silica are used to control tan δ through the interaction of polymer and filler. The chain-end modified polymer was developed in the 1980s by solution polymerization, which allows material engineers to freely design the molecular structure. Solution SBR is representative of this technology. Compared with previously used emulsion SBR, solution SBR has two features:

(i) Polymers of low molecular weight are not generated.
(ii) Chain ends of a polymer can be chemically modified (i.e., chain-end modification).

The first feature reduces the energy loss at chain ends of a polymer because the number of chain ends decreases. The second feature allows good dispersion of carbon black by preventing its aggregation because it can bind with chain ends chemically modified by a functional group. Thus, two kinds of energy loss are suppressed. One is the energy loss due to the change in arrangement of filler particles. This is related to the Payne effect by which the dynamic modulus decreases with increasing strain amplitude. The other is the energy loss related to thermal vibration at free terminals of the polymer chain. Figure 5.17 shows that the temperature dependency of tan δ can be controlled to satisfy compatibility between rolling resistance and braking performance on a wet road using the chain-end modified polymer (solid line) rather than the previous polymer (dotted line).

5.4.3   Silica-Loaded Tire

Tires with silica-loaded tread appeared in the 1990s as discussed in Sect. 5.1.3. If silica is simply mixed to a rubber, silica particles agglomerate to form a particle–particle network. This is because the hydrophilic property of silica surface makes it difficult for silica to homogeneously disperse in a rubber which is hydrophobic. The inhomogeneous dispersion of silica particles increases the hysteresis loss of rubber. In other words, silica-to-silica interaction is much larger than silica-to-rubber one. Hence, utilization of various silane coupling agents has been developed to promote the dispersion of particulate silica homogeneously in the rubbery matrix. The agent binds with the silica surface in a kneading process and makes silica hydrophobic for its homogeneous dispersion. Furthermore, it forms a chemical bond with rubber in the vulcanization process to reinforce the rubber. Because both the dispersion and reinforcing property of silica are improved simultaneously, silica is now applied to a tire in combination with a suitable coupling agent.

The mechanism behind the tan δ being controlled by silica is the same as that of carbon-black-loaded rubber with a chain-end modified polymer, but the interaction of silica with rubber is greater than that of carbon black. Hence, compared with carbon-black-loaded rubber, silica-loaded rubber effectively decreases tan δ at the temperature of 50 °C related to rolling resistance without changing tan δ at the temperature of 0 °C related to braking performance (skid resistance) on a wet road. Figure 5.19 shows the measurement of rolling resistance and braking distance on a wet road for different compounding ratios between carbon black and silica. The rolling resistance is up to 15% less for silica-loaded rubber than for carbon-black-loaded rubber [38].

As a method of further improving the dispersion and reinforcing property of silica, various chemical agents that improve dispersion and a chemically modified rubber for silica loading have been developed. For example, tertiary amine was proposed as an agent for better dispersion: Tertiary amine improved the dispersion of silica by strongly interacting with the nitrogen atom in the silanol group of the silica surface, while simultaneously keeping affinity with rubber at the alkyl

Fig. 5.19 Measurement of the rolling resistance and braking distance on a wet road for different compounding ratios between carbon black and silica (from Ref. [38])

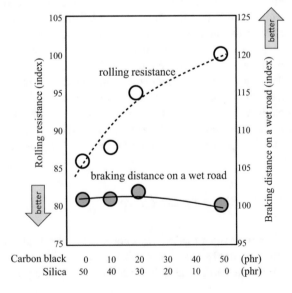

group. This new dispersion-improving agent was reported to reduce rolling resistance by 6% without reducing braking performance on a wet road [39]. Examples of rubber modification for silica include amine–amine modification, alkoxysilane modification, and amine–alkoxysilane modification [39].

The tire and rubber companies have improved the temperature dependency of tan δ of tread rubber over more than 30 years since the 1980s to satisfy the compromising conditions between low rolling resistance and braking performance on a wet road. However, the current tire having a low rolling resistance faces the problem that it does not allow the fuel efficiency of a vehicle to be improved in winter compared with summer [40]. This is because the tread rubber tan δ of low rolling resistance tires is designed to increase tan δ at 0 °C for better braking performance on a wet road while decreasing tan δ at a temperature of 50 °C for lower rolling resistance. The remaining task is to realize the compromise of lower rolling resistance and better braking performance on a wet road without relying on the temperature dependency of tan δ.

5.5   Future of Automobile Tires

5.5.1   Social and Technical Environments Around Tire

Automotive companies are currently manufacturing or developing several types of automobiles, that is, hybrid electric vehicle (HEV), plug-in hybrid electric vehicle (PHEV), electric vehicle (EV), fuel-cell vehicle (FCV). They are supposed to realize a higher mileage and lower (hopefully zero) emission of carbon dioxide, in response to environmental issues such as global warming and the steep rise of fuel prices. The contribution of the rolling resistance of tires to the fuel efficiency of a vehicle with an internal combustion engine depends on the traveling mode and is estimated to be from 5 to 30% for a passenger car and from 15 to 40% for a truck/bus [41]. Meanwhile, the contribution of the rolling resistance of tires to the fuel efficiency of a HEV/PHEV/EV/FCV is larger than that for a vehicle with an internal combustion engine, owing to smaller energy loss and energy regeneration. The Japan Automobile Tyre Manufacturers Association (JATMA) conducted a life cycle assessment (LCA) of a tire by dividing the life cycle into five stages: raw material, manufacturing, distribution, use, and after use (disposal/recycling). As shown in Fig. 5.20, the contribution of the use stage to the generation of carbon dioxide is more than 85% in the LCA of passenger-vehicle tire and truck/bus tire. Hence, the reduction of rolling resistance is more important than energy efficiency in the stages of raw material and manufacturing [39].

From the view of the safety and environment, the European Union began regulating tires in terms of rolling resistance, noise, and braking distance on wet roads in 2012. Japan will introduce the same legislation in 2018. In 2007, the USA made the installation of a TPMS mandatory that monitors the air pressure of a tire to

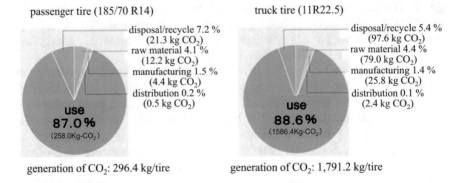

passenger tire (185/70 R14)   truck tire (11R22.5)

generation of CO_2: 296.4 kg/tire   generation of CO_2: 1,791.2 kg/tire

Fig. 5.20 Greenhouse gas emission by tire lifecycle stage (converted to CO_2) (from Ref. [39])

reduce the likelihood of tire failure and avoid the increase in rolling resistance due to the low pressurization of tires. The European Union and Korea introduced similar legislation in 2012 and 2013, respectively, and Japan and China are planning to introduce such legislation, too. This trend may encourage the wide use of run-flat tires, which makes it necessary to install a TMPS.

Some local governments in Japan have conducted demonstration tests of infrastructure of intelligent transport systems such as road-to-vehicle communication and pedestrian–vehicle communication systems from the viewpoint of safety. In a long run, it is expected that the development of intelligent transport systems will expand and that technology for safe driving assistance and more or less autonomous driving will become practical. However, the establishment of these new technologies may need some more time. If such an infrastructure becomes commonplace, traffic accidents are hopefully predicted to decrease, even when the improvement of tire performance in terms of braking and maneuverability is delayed. However, the importance of the braking performance and maneuverability of a tire may not be neglected in order to cope with the ever increasing number of automobiles and their constant and world-wide spreading for a while. Tires may differ from currently available pneumatic tires, in order to satisfy various and many requests from concerns on safety and environmental performances, e.g., lower noise, appropriate grip, good skid resistance, low rolling resistance, appropriate wear (balancing with that of road surface), and lots of other issues relevant to rubber tires.

5.5.2 Tire Technology in the Twenty-First Century

(1) Relationships between Tire Performance and Tire Design Elements

The future trend of tires influenced by and in accordance with the social environment may be foretold by the relationships between tire performance and tire design elements in Table 5.4, which is a detailed version of Table 5.2 [42]. Tire

Table 5.4 Relationships between tire performances and tire design elements

Design parameter			Safety		Amenity	Environment
Design element	Design parameter	Direction of change	Brake	Maneuverability	Riding comfort	Noise
Shape dimension	Diameter	Large	A	A	C	A
	Width	Wide	A	A	C	B
Construction	Belt	Stiff	A	A	B	A
	Side wall	Stiff	A	A	B	A
Pattern	Groove volume	Big	A: WET B: DRY	B	C	B
Material (tread rubber)	Hardness	Stiff	A	A	B	B
	tan δ	Large	A	C	A	C
Inflation pressure		High	B	Appropriate	B	C

Design parameter			Environment		
Design element	Design parameter	Direction of change	Rolling resistance	Wear	Weight
Shape dimension	Diameter	Large	A	A	B
	Width	Wide	A	A	B
Construction	Belt	Stiff	A	A	C
	Side wall	Stiff	B	C	C
Pattern	Groove volume	Big	A:PS B:TB	B	A
Material (tread rubber)	Hardness	Stiff	Appropriate	A	C
	tan δ	Large	B	A	C
Inflation pressure		High	Appropriate	Appropriate	C

From Ref. [42]

Performance of a tire: *A* better; *B* worse; *C* little change; appropriate, exists appropriate value, *PS* passenger-vehicle tire; *TB* truck/bus tire

performance is classified into safety (braking and maneuverability), amenity (maneuverability and riding comfort), and environmental (tire noise, rolling resistance, wear, and weight) performance. Although tire design is usually discussed in terms of four elements (shape, construction, pattern, and material) as shown in Table 5.2, the dimensions of a tire and air pressure are further added as design elements in Table 5.4. When one design parameter is slightly changed without changing other design parameters, "A" indicates that the performance improves, "C" indicates that the performance changes little, "B" indicates that the performance worsens, and the term "Appropriate" indicates that the appropriate value exists between upper and lower ends of the design parameter.

The type of symbol in Table 5.4 is mainly determined according to simple tire theories. Maneuverability is evaluated in terms of cornering stiffness according to Fiala's theory [43], braking performance is evaluated in terms of the braking stiffness [44], rolling resistance is evaluated in terms of the summation of compressive energy loss and shear energy loss in the tread [45], wear is evaluated in terms of wear energy according to Schallamach's theory [46, 47], riding comfort is evaluated in terms of the enveloping property of a tire rolling over a cleat [44, 48] and tire noise is evaluated according to Nakajima's theory on the pitch noise of a tire [25].

Because there is a mix of symbols "A" and "B" for each design element in Table 5.4, we see the difficulty in satisfying many aspects of performance at the same time in tire design. For example, if we decrease tire noise by reducing the tire width under the design parameter of shape and dimension, the wear, maneuverability, and rolling resistance will be worse. If we decrease the tire noise by reducing the groove volume, the braking performance on a wet road and rolling resistance of the passenger-vehicle tire will be worse. Note that the tire diameter is a specific parameter in all design elements, with all performances except weight improving if the tire diameter is increased.

(2) **Future Vision of Tires**

Considering the social environment in particular, future passenger-vehicle tires may be classified into five categories:

 (i)   inexpensive pneumatic tires;
 (ii) expensive pneumatic tires with a low aspect ratio that have the distinguished specific ability to run flat, good maneuverability, or riding comfort;
 (iii) downsized pneumatic tires having high air pressure and drastically improved rolling resistance and tire noise;
 (iv) airless tires for small and low-speed mobility vehicles;
 (v)   intelligent tires.

The first two categories are a continuation of the trend of current tire development. The third category is a downsized pneumatic tire in which noise and rolling resistance are selectively and drastically improved, since the improvement of both environmental performances is realized by tire among three elements for mobility (road, vehicle, and tire elements). The downsized tire is a pneumatic tire with a small cross-sectional area and high air pressure shown in the middle of Fig. 5.21, and it can be designed by using Table 5.4 for the relationships between tire performance and tire design elements. First, tire noise can be reduced by developing a tire narrower than the control tire. However, a narrow tire cannot support the same load as the control tire, and higher air pressure is required. The rolling resistance is reduced by both the higher air pressure and the decrease in rubber volume. However, the wear of a downsized tire inevitably becomes worse owing to the smaller contact width, the maneuverability and braking performance becomes worse owing to the shorter contact length, and the riding comfort becomes worse owing to the higher spring constant resulting from higher air pressure.

One way of improving these worsened performance areas is to increase the diameter of a downsized tire as shown in Table 5.4 and on the right of Fig. 5.21. Because maneuverability, braking performance, and wear are improved by a longer contact length, some of the worsened performances of a downsized tire will be mitigated by a larger diameter. Kuwayama et al. [49] compared performance

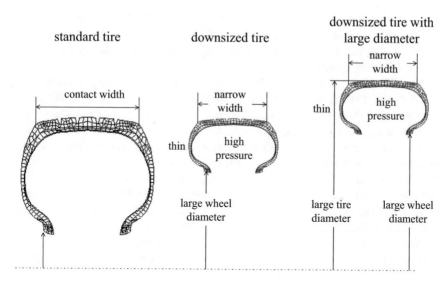

Fig. 5.21 Downsized tire (from Ref. [42])

between an ordinary tire and a downsized tire with high pressure and large diameter. The size and air pressure of the former tire were 175/65R15 and 220 kPa, while the size and air pressure of the latter tire were 155/55R19 and 320 kPa. They showed that the maneuverability and braking performance were improved by a larger diameter. A downsized tire with a large diameter was adopted as a standard tire for an electric vehicle made by BMW.

Another way of improving the worsened performances of a downsized tire is to use all design elements for mobility; i.e., not only design elements of a tire but also design elements of the road and vehicle. The three types of element for mobility contribute to the performance of a vehicle at various levels. If each element improves a certain aspect of mobility performance, each aspect of performance may become satisfactory. For example, the worsened performances of a downsized tire are braking performance, maneuverability, and riding comfort. The braking performance may be improved by an intelligent transport system (e.g., a road-to-vehicle or pedestrian-to-vehicle communication system), the use of which is expected to be widespread in the future. Furthermore, braking performance may be improved by a road with water-permeable pavement. Maneuverability may be improved by control technology such as electric stability control or autonomous driving. The remaining reduced performance of riding comfort may be improved by the suspension of a vehicle.

The fourth category is an airless tire without air pressure made by injection molding. The airless tire might give the same impact to the tire industry as the steel belted radial tire, if the remaining problems discussed later were overcome. Because there will be three simultaneous technological changes in terms of the material, construction, and tire building of the airless tire, which is similar to the case of the development of the steel belted radial tire in Sect. 5.1.3. New tire manufacturers will be able to enter the tire industry easily because of the greatly simplified manufacturing process, and then the price of the airless tire will rapidly decrease after it enters widespread use.

Michelin proposed the airless tire "Tweel" in 2005 [50]. Injection-molded tires called casting tires have been studied by many tire companies since the 1960s, but they were pneumatic tires. Although Tweel may not be made by employing a new manufacturing method, it is as great an invention as the airless tire. The merits of Tweel are not only its simple manufacturing process of injection molding, but also its safety in the event of a puncture and its use of recyclable polyurethane for a skeleton member. However, Tweel currently faces problems relating to rolling resistance, wind noise, and axle vibration due to the spoke-type structure [51]. Other problems of the airless tire may be the deterioration and creep of the polyurethane material, fatigue fracture of spokes due to buckling deformation in the contact area between tire and road, and irregular wear and degradation of maneuverability due to an uneven pressure distribution or higher contact pressure at the spoke joints. It may take time to solve these problems. Hence, the airless tire may be first enter the market for military vehicles or use in low-weight/speed vehicles, such as mobility devices for elderly people, whose numbers are increasing, rental vehicles for tourist destinations, and small vehicles for congested urban areas. The puncture resistance and maintenance-free nature of airless tires will be advantageous in these markets, and the aforementioned problems do not apply.

The fifth category is an intelligent tire that improves safety and maneuverability through the monitoring of tire behavior. The technology used to monitor the tire state by gluing a sensor inside a tire has been practically applied to the tires of earthmovers. Additionally, the aforementioned TPMS is an intelligent tire in a broad sense. The other example is the intelligent tire in which the friction coefficient of the road surface can be sensed by a strain gauge or accelerometer glued inside a tire and the change in road conditions can be instantaneously transmitted to the vehicle from the sensor [52]. Because a tire is the only point of contact between a vehicle and road, it can monitor road surface conditions earlier than other parts of a vehicle. However, many problems also remain for the intelligent tire; e.g., problems relating to the method of gluing a sensor inside a tire, the durability of sensors, communication means, and battery life.

(3) **Future Tire Technology**

The trend of future tire technology may be understood by the four guiding principles:

(i)   change of the scale of observation from the macroscopic to the microscopic and then the nanoscopic level;
(ii) integration and fusion of technologies;
(iii) sustainability;
(iv) individual innovative technologies.

The first trend comes from the reductionism of René Descartes that our under-standing of objects is deepened by changing the scale of observation as shown in Table 5.5. At the macroscopic scale, tire performances such as the rolling resis-tance, wear, and maneuverability are measured. For example, tire performances are related to the macroscopic pressure distribution between the tire and road controlled by the crown shape and other design elements. Tire performances can be further understood according to basic properties such as viscoelasticity, friction, and wear at the microscopic scale. The microscopic properties are related to the microscopic pressure distribution controlled by the interaction between the rubber and micro-scopic road roughness. The microscopic phenomena are understood in terms of the nanoscopic properties of rubber governed by, for example, the polymer and dis-persion of filler. Nanoscopic phenomena are fundamental to tire performances. Hence, we may be able to develop new technology for macroscopic tire perfor-mances by deepening our understanding of phenomena that occur at the microscopic/nanoscopic level and the relationships between properties at different scales, e.g., macro vs. micro and micro vs. nano.

All tire companies focus on the technology of microscale/nanoscale material design because such material is the largest contributor to tire performance and it is in the second and third of the three scales of science given in Table 5.5. Furthermore, the microscale/nanoscale region ranging from 1 nm to 1 μm is par-ticularly meaningful for polymer materials because this region includes various parameters such as the structure of the amorphous polymer, the thickness of the crystal lamella of the crystalline polymer, the microphase separation of the block graft copolymer, the interpenetrating polymer network, and the phase separation of the polymer blend and nanocomposites [53]. Measurements are actively performed in the area of micro-/nanotechnology employing the three-dimensional electron microscope [54–56], X-ray microscope [57], and atomic force microscope [58], and computer simulations are used to visualize the interaction between the polymer and filler [59, 60]. Additionally, nanoscopic mechanical properties are mapped by measuring the reaction of an atomic force microscope probe pushed against rubber [58], and the contribution of hysteresis and cohesion to friction have been estimated according to the microscopic behavior on the contact interface between rubber and

Table 5.5 Three scales in the technology of tires

Scale	Tire performance	Contact pressure between tire and road
Macro	Rolling resistance, wear, maneuverability, etc.	Macroscopic pressure distribution controlled by crown shape and other design elements
Micro	Viscoelasticity, friction, wear, etc.	Microscopic pressure distribution controlled by interaction between rubber and microscopic road roughness
Nano	Nanoscopic properties of rubber controlled by polymer, dispersion of filler, etc.	

Prepared by Y. Nakajima

glass that has been observed by applying high-intensity X-rays of synchrotron radiation at SPring-8 [57]. Hereafter, it is expected that technology will be optimized to control the interaction between conventional polymer and filler and that engineers will then be able to design material by considering the hierarchical structure from the microscale/nanoscale to the macroscale in the final stage.

The second trend is the integration and fusion of technologies. As discussed in Sect. 5.3.1, the technology for the design of the tire shape has been established and there are few remaining areas where this technology can be improved. Hence, the technology for the integrated design of shape and construction will hereafter be developed to allow us to optimize these features simultaneously. Furthermore, if the three elements for mobility (road, vehicle, and tire elements) are integrated in vehicle design, the important aspects of tire research will be rolling resistance, noise, and wear because the tire contributes greatly to these performance areas.

Tires have not benefitted from the technological progress in electronics and information technology because it is difficult to adapt these technologies to tires. If new value is added to a tire by the fusion between tire and electronics or information technology, there is the possibility that a completely new tire will be created.

The third trend is the sustainability. The technology development to utilize natural resources rather than fossil resources is essential for sustainable society. Natural rubber is the important natural resource for tires, but it has become an unevenly distributed resource. Because the natural rubber plantations in South America have been destroyed by South American leaf blight, more than 90% of natural rubber is now produced in Southeast Asia. Furthermore, there is the risk of diseases other than South American leaf blight emerging. To mitigate such risks, rubber plantations need to become more distributed and various methods of rubber production need to be studied. For example, plants other than *Hevea brasiliensis*, such as Guayule and Rubber dandelion (see Sect. 3.4.2 for them), are being cultivated and the yield of rubber may be improved through the gene recombination of *H. brasiliensis* [61–63].

The recycling of tires is also important for sustainable society because rubber tires play a great role in transportation. Currently, pneumatic motor vehicle tires cannot be recycled because the wear they endure actually reverses the vulcanization process, and technological solutions to this problem have yet to be found. Thus, recycling is now limited to bicycle tires. However, although they cannot be recycled, motor vehicle tires have been reused since early days. Reuse of truck/bus tires has had particular success through retreading worn-out tires [63–66]. From the point of view of material recycling, the retread process of replacing the tread on worn tires will become widespread in all tire categories, though we did not discussed it in this chapter.

The fourth trend is the individual innovative technologies such as the automatic manufacturing system discussed in Sect. 5.1.3 and the airless tire.

Remark 4 Seismic Isolation Device Using Rubber [67, 68]
The technology to reduce the effect of seismic vibration on a building is classified into vibration resistance, vibration control and vibration isolation. The vibration resistance

is to make a building durable against an earthquake by increasing the stiffness, the vibration control is to absorb the seismic vibration by placing dampers under a building, and the vibration isolation is to attenuate the vibration of a building by the seismic isolation device such as the rubber-metal laminated bearing (often called the laminated rubber bearing) which isolates a building from the quake of the ground.

History of the application of the device for vibration isolation is relatively new. It has been said that the non-laminated rubber bearing was first used for the building at Skopje in Macedonia in 1969, but it did not adequately isolate the vibration. The device has to adequately attenuate the quake in the horizontal direction as well as support the load of building, in order to realize the vibration isolation. These functions have been realized by the laminated rubber bearing where steel plate and thin rubber layer are laminated alternately. The structure of the laminated rubber bearing not only suppresses the barrel deformation of rubber between steel plates caused by the load but also provides the small horizontal rigidity which is a few thousandth of the vertical rigidity.

The cylindrical shape is generally used in the laminated rubber bearing for a building. The load capacity of the laminated rubber bearing is proportional to the cross-sectional area of the cylinder, while the horizontal load capacity is proportional to a diameter of the cylinder. The laminated rubber bearing isolates the building from the quake of the ground, and the acceleration and displacement of the building is found to decrease by from 60% to 80%. The laminated rubber bearing prefers natural rubber for its property of excellent linearity and a stable restoring force. However, it inherently has a low damping factor and the building with these devices may continue to quake even after the quake of the ground is stopped. Hence, separate dampers are required to give the viscous property to the structure. Examples of the damper are the oil damper using the damping function of highly viscous fluid, a lead plug damper using the hysteresis curve of the elastic-plastic material which is embedded at the center of a laminated natural rubber bearing, and a stainless steel slide plate using the friction force between the plate and a building. Meanwhile, the highly damping rubber bearing has been developed to increase the damping property of natural rubber by loading highly viscous rubber, resin or silica particles to the natural rubber.

When the laminated rubber bearing is applied to a detached house with a light weight, the deformation capability may not be adequate to support the load of the house, or the acceleration of an earthquake may not be attenuated adequately owing to the small diameter of the bearing. Therefore, the device using the steel slide plate instead of the laminated rubber bearing has been developed for the detached house. The laminated rubber bearing has the problem that it may not adequately isolate the vertical quake due to the large vertical stiffness when an earthquake occurs just under the structure. This problem has been solved by the technology of the three-dimensional bearing.

It has been expected that the laminated rubber bearing is possibly applied not only to new buildings but also to existing buildings which may be vulnerable to an earthquake. The latter application, now called the base-isolation retrofit, tends to increase for existing buildings which store important cultural assets or which are themselves important cultural assets. The base-isolation retrofit is of value for preservation and even for survival of culturally important assets in the areas where have a high incidence of earthquake.

References

1. 1. Japan Automotive Tyre Manufactures Association (JATMA) home page, *History of tire in 5000 Years*. http://www.jatma.or.jp/tyre5000/. Accessed 6 Oct 2015
2. T. Watanabe, *Story of Tire* (Japanese Standards Association, Tokyo, 2002) [In Japanese]
3. E. Tompkins, *The History of the Pneumatic Tyre* (Eastland Press, London, 1981)
4. J.P. Norbye, *The Michelin Magic* (TAB Books, Blue Ridge Summit, PA, 1982)
5. H.R. Lottman, *The Michelin Men Driving an Empire* (I. B, Tauris, Paris, 2003)
6. T. Maniwa, *Knowledge and Property of Automotive Tires* (Sankaido, Tokyo, 1979) [In Japanese]
7. S. Kohjiya, *Natural Rubber: From the Odyssey of the Hevea Tree to the Transportation Age* (Smithers Rapra, Shrewsbury, 2015), pp. 149–151
8. W.J. Woehrle, Automotive Engineering **71**, September, 71 (1995)
9. B.N. Zimmerman (ed.), *Vignettes from the International Rubber Science Hall of Fame (1958–1988): 36 Major Contributors to Rubber Science, Rubber Division of Am* (Chem. Soc, Akron, 1989)
10. Y. Nakajima, Nippon Gomu Kyokaishi **85**, 178 (2012) [In Japanese]
11. Y. Nakajima, J. Soc. Automotive Eng. Jpn. **67**, 4 (2013) [In Japanese]
12. M.J. French, *The U.S. Tire Industry: A History* (Twayne Publishers, Boston, 1991)
13. Y. Hirata, H. Kondo, Y. Ozawa, in *Chemistry, Manufacture and Applications of Natural Rubber*, ed. by S. Kohjiya, Y. Ikeda (Woodhead/Elsevier, Cambridge, 2014), Chap. 12
14. H.F. Schippel, Ind. Eng. Chem. **15**, 1121 (1923)
15. R.B. Day, J.F. Purdy, *Goodyear Research Report* (Goodyear Tire & Rubber Co., Akron, 1928)
16. J.F. Purdy, *Mathematics Underlying the Design of Pneumatic Tires* (Hiney Printing Co, Ann Arbor, 1963)
17. F. Böhm, Automobeltechnische Zeitschrift **69**, 255 (1967)
18. K. Yamagishi, M. Togashi, S. Furuya, K. Tsukahara, N. Yoshimura, Tire. Sci. Technol. **15**, 3 (1987)
19. H. Ogawa, S. Furuya, H. Koseki, H. Iida, K. Sato, K. Yamagishi, Tire. Sci. Technol. **18**, 236 (1990)
20. Y. Nakajima, T. Kamegawa, A. Abe, Tire. Sci. Technol. **24**, 184 (1996)
21. Y. Nakajima, H. Kadowaki, T. Kamegawa, K. Ueno, Tire. Sci. Technol. **27**, 62 (1999)
22. T. Akasaka, in *Textile Structure Composites,* ed. by T-W. Chou, F.K. Ko (Elsevier, New York, 1989), Chap. 9
23. Y. Nakajima, *SAE Paper*, No. 960997 (1996)
24. A. Abe, T. Kamegawa, Y. Nakajima, Optim. Eng. **5**, 77 (2004)
25. Y. Nakajima, J. Vibration and Acoustics **125**, 252 (2003)
26. E. Seta, T. Kamegawa, Y. Nakajima, H. Ogawa, Tire. Sci. Technol. **28**, 140 (2000)
27. Y. Nakajima, Int. J. Automot. Technol. **1**, 26 (2000)
28. E. Seta, T. Kamegawa, Y. Nakajima, Tire Sci. Technol. **31**, 173 (2003)
29. JATMA, *Tyre Industry in Japan* (Japan Automotive Tyre Manufactures Association, Tokyo, 2015) [In Japanese]
30. Bridgestone (ed.), *Pneumatic Tire Technology* (Sankaido, Tokyo, 2006) [In Japanese]
31. K. Akutagawa, Y. Ozawa, H. Yamada, T. Hamada, Nippon Gomu Kyokaishi **80**, 394 (2007) [In Japanese]

32. H. Takino, R. Nakayama, Y. Yamada, S. Kohjiya, T. Matsuo, Rubber Chem. Technol. **70**, 584 (1997)
33. H. Takino, H. Takahashi, K. Yamano, S. Kohjiya, Tire. Sci. Technol. **26**, 241 (1998)
34. H. Takino, N. Isobe, S. Kohjiya, Tire Sci. Technol. **26**, 258 (1998)
35. S. Futamura, Tire Sci. Technol. **18**, 2 (1990)
36. H. Takino, S. Iwama, Y. Yamada, S. Kohjiya, Rubber Chem. Technol. **70**, 15 (1997)
37. H. Kaidou, F. Yatsuyamagi, Nippon Gomu Kyokaishi **74**, 159 (2001) [In Japanese]
38. M. Matsuura, Nippon Gomu Kyokaishi **71**, 583 (1998) [In Japanese]
39. Y. Ozawa, K. Akutagawa, K. Yanagisawa, Y. Hirata, Nippon Gomu Kyokaishi **77**, 219 (2004) [In Japanese]
40. H. Suzuki, K. Yamaguchi, I. Inoue, *Proceedings of JSAE Technical Meeting*, No. 20125688 (2012)
41. J. Barrand, J. Bokar, *SAE Paper*, No. 2008-01-0154 (2008)
42. Y. Nakajima, Noise Control **39**, 50 (2015)
43. E. Fiala, VDI Zeitschrift **96**, 973 (1954)
44. H. Sakai, *Taiya Kougaku (Tire Engineering)* (Guranpuri Shuppan, Tokyo, 1987). [In Japanese]
45. T.B. Rhyne, S.M. Cron, Tire. Sci. Technol. **40**, 220 (2012)
46. A. Schallamach, D.M. Turner, Wear **3**, 1 (1960)
47. Y. Nakajima, Nippon Gomu Kyokaishi **88**, 31 (2015) [In Japanese]
48. S. Gong, *A Study of In-plane Dynamics of Tires*, Ph.D. Thesis, Delft University of Technology, Delft (1993)
49. I. Kuwayama, H. Matsumoto, H. Heguri, *JSAE Paper*, No. 90–20135154 (2013)
50. Michelin Home Page, http://www.michelintweel.com/. Accessed 6 Oct 2015
51. W. Rutherford, S. Bezgam, A. Proddaturi, L. Thompson, J.C. Ziegert, T.B. Rhyne, S.M. Cron, Tire Sci. Technol. **38**, 246 (2010)
52. H. Morinaga, Development of sensing algorithm for intelligent tire, *Tire Technol. Expo*, Cologne (2006)
53. T. Nishi, Nippon Gomu Kyokaishi **76**, 358 (2003) [In Japanese]
54. H. Jinnai, Nippon Gomu Kyokaishi **76**, 384 (2003) [In Japanese]
55. Y. Ikeda, A. Katoh, J. Shimanuki, S. Kohjiya, Macromol. Rapid Commun. **25**, 1186 (2004)
56. S. Kohjiya, A. Katoh, Y. Ikeda, Prog. Polym. Sci. **33**, 979 (2008)
57. N. Amino, Nippon Gomu Kyokaishi **85**, 332 (2012) [In Japanese]
58. S. Fujinami, K. Nakajima, T. Nishi, Nippon Gomu Kyokaishi **84**, 171 (2011) [In Japanese]
59. Y. Masubuchi, Nippon Gomu Kyokaishi **82**, 459 (2009) [In Japanese]
60. Y. Tomita, Nippon Gomu Kyokaishi **82**, 464 (2009) [In Japanese]
61. M. Hojo, Y. Fukushima, T. Sato, N. Watanabe, Y. Ozawa, Nippon Gomu Kyokaishi **86**, 169 (2013) [In Japanese]
62. Y. Ikeda, S. Kohjiya, in *Sustainable Development: Processes, Challenges and Prospects*, ed. by D. Reyes (Nova Science Publishers, New York, 2015), Chap. 3
63. S. Kohjiya, *Natural Rubber: From the Odyssey of the Hevea Tree to the Transportation Age* (Smithers Rapra, Shrewsbury, 2015), pp. 250–261
64. A.I. Isayev, in *Chemistry, Manufacture and Applications of Natural Rubber*, ed. by S Kohjiya, Y Ikeda (Woodhead/Elsevier, Cambridge, 2014), Chap. 16
65. Y. Ikeda, in *Chemistry, Manufacture and Applications of Natural Rubber*, ed. by S Kohjiya, Y Ikeda (Woodhead/Elsevier, Cambridge, 2014), Chap. 17
66. M. Mccoy, Chem. Eng. News, April 20, **93**, 16 (2015)
67. The Japan Society of Seismic Isolation, *Introduction to Seismic Isolation Laminated Rubber* (Ohmsha, Tokyo, 1997) [In Japanese]
68. J. M. Kelly, *Earthquake-Resistant Design with Rubber*, 2nd ed. (Springer, London, 1997)

Chapter 6
Rubber Science and Technics Toward the Next Century: A Prospective View

6.1 Sustainability and Rubber

Sustainable development (SD) of the present human beings on the earth is the most important issue of this century. In other words, how we can develop ourselves without further damaging all the natural existences is the question we are facing now. For a moment, we have to be practical in living a life in accordance with SD. In other words, SD may be an ideology itself, but it is now to be practiced in our everyday life. Otherwise, the next century could be the last one for us to live on the earth—is one of the possible consequences by scientific thought or by a priori thinking as has been suggested by the well-known THE CLUB OF ROMES report which has been disseminated since 1972 [1].

The basic idea of SD is a harmony between the activity of human beings and the nature. This idea is very old, since nature is much older than human beings. We were totally under the control of natural incidents in the past, and human beings are much older than science. Nature was active, and we were passive then [2, 3]. On the other hands, technics was as old as human beings. Note that "technology" has been used instead of "technics" in the past, but technology had better be a science on technics as described in Chap. 1. Hence, technology is a branch of science, and rubber science is a sub-branch of technology. In other words, we have to distinguish technology from technics itself for the usage in this chapter. The human beings began to use a simple tool made of wood and/or stone (the origin of technics) in labor and use language in the community or in the families. Both technics and language have been the essential tools of human beings for their evolution as well as their survival on the earth.

In spite of using tools and language, our passiveness against the nature might have been ubiquitous before the civilization, i.e., until the beginning of agriculture about ten thousand years ago, which was followed by the increasing utilization of ironware [4–8]. When hunting-and-gathering economy had been predominant, people were forced to move, chasing the catch and wild eatable plants. Agricultural

© Springer Nature Singapore Pte Ltd. 2018
Y. Ikeda et al., *Rubber Science*, https://doi.org/10.1007/978-981-10-2938-7_6

activity, even with its primitive technics, enabled us to settle down for a longer time at a place where flat lands and water were available. After settling down to engage in agricultural activities, people modified the encircling nature by using hand tools, to result in the formation of environment instead of the genuine nature. That is, civilization began with agriculture, the development was accelerated by using iron tools, and the process of converting nature to environment has been accelerated since then [2, 4–6, 9, 10] (see Fig. 1.3 in Chap. 1). The conversion of nature to environment was a result of their struggling for survival and might not be their intention at all then.

Incessant progress of technics was observed all through the classical (when the Greek and the Roman were ruling) and the medieval periods. Even during the early middle age, when still referred to as a "dark" age, "the arable lands of northern Europe were still being extended piecemeal, generation by generation, from hard-worn forest" was the case [7]. The farmers' efforts were supported by medieval technics, which were originated from the three sources; the classical ones of Greeks and Romans, Islamic and Asian ones, together with its own creativity, of course [11].

Then, the Renaissance foretold a new stage in the relationship of us with nature. Leonardo da Vinci (1452–1519) was not educated at schools where medieval Scholasticism was predominant, and had not much interaction with the contemporary Humanists. Thus, he managed to face nature without any prejudice but only with his open eyes [12]. All his multitalented achievements including scientific ones (inclusive of engineering arenas), suggested a future trend of human activities even though he was not a scientist in a modern sense. A little later, Francis Bacon (1561–1626) had authored "*The Advancement of Learning*" in 1605 and "*Novum Organum*" in 1625. He himself did not do any scientific research activity, but his proposal of how to approach or work on nature by now called "the induction method" stimulated much to give rise to the future scientific activity. F. Bacon owed much to Roger Bacon (ca. 1214–after 1292) is to be added. R. Bacon was a Franciscan and a forerunner in terms of opposition to the Scholasticism. Interestingly however, F. Bacon did not study the scientific achievements by the contemporary scientists such as Nicolaus Copernicus (1473–1543), Johannes Kepler (1571–1630), and Galileo Galilei (1564–1642), which might be the reason why his philosophy of science lacked the recognition of mathematics and its importance in developing science [12].

Through the Scientific Revolution in the sixteenth and seventeenth centuries [9, 13, 14], the modern science showed its presence by elucidating the secrets of nature, followed by the progress of related technics. In the arena of philosophy then, human beings often confronted with the nature, and a belief that we might be able to control nature by means of science and technics became more and more popular [9, 15–17]. This novel trend toward developing civilized society was decisively promoted physically by a quantum jump, which was due to Industrial Revolution [18, 19]. In fact, the industrial revolution enabled the quantum jump of technics, among which the development of steam engines is to be noted in particular, followed by evolution of technology (science of technics) and remarkable progress of science in the nineteenth century. These powerful trends greatly helped the idea of omnipotence of science and technics to be spread all over the world even among ordinary people [16, 20–26] (see also Sect. 1.3.3).

Historically, start of the jump was achieved by the steam engines [27]: Origin of the word "engineer" or "engineering" can obviously be "engine." Note that the industrial revolution had been started by the use of hydraulic power as well as the use of domesticated animals and the wind power. The two had been practiced and spread during the medieval periods, while introduction of hydropower was relatively new. Therefore, "James Watt (1736–1819) was the first engineer" is uniquely a correct statement, with the constant support of Matthew Boulton (1728–1809). Watt's uniqueness was his deep knowledge on chemistry as well as mechanical engineering. He was actually at the engineering workshop of the University of Glasgow, within the confines of the College [27].

The steam engine sent a steamer to the sea, and a steam locomotive to the land early in the nineteenth century, both of which initiated the Steam Age, the initial step of the Transportation Age. It was further accelerated by the inventions of an automobile and an aircraft in the late nineteenth and early twentieth centuries, respectively. The Transportation Age has necessarily been promoting delocalization and globalization. Or, the delocalization trend has promoted the development of physical means of transportation? It is a kind of "chicken or egg" discussion. Above all, the significance of automobiles equipped with pneumatic rubber tires is worth being mentioned from a historical viewpoint [26]: While steam locomotives are destined to move on the iron rails, automobiles afford much more freedom for delocalization and globalization movements, and it has been the nucleus of the transportation network system since the middle of the twentieth century.

As noted in Sect. 3.1, above descriptions on the Scientific Revolution and the Industrial Revolution have to be complemented by the recent evolutionary idea in describing development or historical progresses (may not exactly be in line with the evolution theory in biological arena), that is, the development of science and technics is fundamentally via gradual changes, not by a quantum leap which is too often named "revolution." For literatures of specifically on the evolution side, see the references cited in Sect. 3.1, but even those cited here (Refs. [4, 8, 12, 22, 25, 26]) are not against an evolutionary opinion.

From the birth of human beings, transportation has been essential for them, namely they originally had separated from the ape by walking on foot touching the ground. Transporting or moving in person carrying a luggage, absolutely not alone but in a group, was also an important issue for them to work and to live together in a community, and ultimately to communicate together by inventing the language. It is interesting to note that the word "communicate" has originally the same meaning as transport: Communication is exactly a transportation of information. Commuter is still now nearly equal to transporter, too. Thus, transportation had always been humane, but the division or specialization of work influenced on transportation. Even in the civilized world, only a small number of rich people had afforded to have transporters, e.g., workers (most probably, slaves) with domesticated animals for carrying purposes. In the army, the cavalry was limited to the noblemen, and the horse carriage was only for the nobility. Queen Victoria in the nineteenth century was transported in a horse carriage equipped with iron tires (iron was the encircling cover of a wooden wheel) which, you can imagine, was very uncomfortable ride due to strong jolting and high noise. Only at her last years, she might have enjoyed somewhat comfortable drive on a carriage with solid rubber tires [26].

Such situation had lasted at least for several thousand years. Only as late as the beginning of twentieth century, automobiles equipped with pneumatic rubber tires rapidly expelled the horse carriages from the road. Typically, Ford Model T (the first one was marketed in 1908) enabled its spread among thousands of farmhouse due to its low price or in exchange of the horse carriages as well as among citizens. American farmers once called a Ford T car as a "horseless carriage" before the widespread availability of automobiles in the countryside [26, 28]. Combined with trains, steamships, and aircrafts, automobiles have been central in the transportation networks of the modern society, and pioneering to spread to every corner of the world.

Originally speaking, the word, logistic is used in philosophy, and one derivative word from it, "logistics" has been for military use: Behind the frontline of battle fields, the units in charge of logistics are to secure the supply of food and militaryware and transport them to the battle field in need of them. This word has recently been more or less demilitarized and is generally used for administering or taking care of supply, storage, and transportation of necessary items, even including the dispatch of necessary number of people with special talents or skills. Most recently, it has further widened to mean, in particular, the system and the technics for general management of infrastructure [29, 30], which may be suggesting the maturing of the transportation society, at least in developed countries. Every time when a natural disaster (regrettably, including the accidents of the Fukushima Daiichi nuclear power plant in 2011 [31] and of the Chernobyl nuclear power plant in 1986) occurs, the most important issue in politics and economics is the rapid recovering of the infrastructure of the crashed area. Among many issues, the recovery of the road system to make trucks and automobiles navigable is always the most urgent one from the viewpoint of logistics. The Roman's words, "All roads lead to Rome" are renewed in the modern society. On the Roman Empire roads, horse carriages fully loaded with equipped troops and militaryware rushed to the front line, with so much vibration and big noise, on the stone-paved roads. In the modern society, however, big trucks and automobiles equipped with rubber tires rush on the roads much more smoothly and efficiently, once the repair of roads is completed.

6.2   Automobiles and Transportation-Network Society

In this transportation society, how to understand and practice sustainable development (SD) in our daily life is to be discussed. On this occasion, firstly definition of SD is given just to reconfirm. World Commission on Environment and

Development (WCED), an organization under United Nations [32], announced SD is

Development that meets the needs of the present without compromising the ability of future generations to meet their own needs (WCED, 1987)

This thesis is restated by an economist as [33]

An increase in well-being today should not have as its consequences a reduction in well-being tomorrow

In order to better understand these definitions, impact factor S_I is introduced [26, 34]. S_I is a load factor or a parameter evaluating the load to environment. It may be assumed to be proportional with population P and energy consumption E. That is,

$$S_I = P \times (\text{GDP}/P) \times (E/\text{GDP}) \tag{6.1}$$

where GDP is general domestic production, and GDP/P and E/GDP are production per person and energy consumption per GDP, respectively. The former may be proportional with living standard, and the latter may be inversely proportional with energy efficiency. Presently, it is recognized that the increase of atmospheric carbon dioxide (CO_2) is induced by the increasing human activity since the Industrial Revolution, and the increase seems to have resulted in the rapid augmentation of S_I. The Kaya equation has taken this factor into account to give

$$S_I = P \times (\text{GDP}/P) \times (E/\text{GDP}) \times (CO_2/E) \tag{6.2}$$

Further, consideration of all the possible factors may give the following general equation

$$S_I = [P \times (\text{GDP}/P) \times (E/\text{GDP}) \times (CO_2/E)]\left[\sum W_i(t)A_i(E)/E\right] \tag{6.3}$$

where $W_i(t)$ is a weight of ith load factor at time t, and $A_i(E)$ is an energy consuming ith factor. The i is a natural number from one to n (the total number of load factors), and $\sum$ is the summation of them.

In the practical discussion of SD, the factors in the third term of Eq. (6.3) have to be considered in details, how they contribute to the increase of carbon dioxide, which is surely the most important at present, etc. Further, their relations to energy consumption E and to population P are to be under the consideration. Also, any possibility of more terms should not be forgotten. For example, "food mileage" may be an everyday issue from the SD viewpoint [35]. Some food stuffs produced at the opposite side of continent or ocean are nowadays often on the table, and such incidents are becoming more and more a popular scene in our life. In this maturing transportation society, transcontinental or transoceanic transports of fresh and processed foods may be rational even from SD standpoint. However, the import of fresh fruits or vegetables or high-priced meats by air-flights to reach the consumers' tables in a few days is rational? Standing on this question, people's movements aiming "Local consumption of local products" are active in lots of countries now. Such trials covering many other merchandise as well as food stuffs are expected to contribute to the local economy, which ultimately are of use to promote SD worldwide [36].

Here, it is assumed that automobiles have increasingly played the central role since the middle of the twentieth century, and Transportation Age has supposedly become into the second stage of it, to be called as the automobile-centered period. At the first stage, various means of transportation had been playing each function, and none was much predominant among them in terms of their social importance. Even under this division of the period, however, it is notable that the progress in technics of the pneumatic rubber tires was not subject to the division: The basic rubber tire technology matured in the 1980s and/or 1990s. In a sense, rubber tires had apparently provided the automobiles with so good a performance to be used in the everyday life of ordinary people, which has accelerated their advance into the second stage by contributing to the familiarization of automobiles. In terms of speed, the possible movements of automobiles are too high, and the weights of them are too heavy to be handled in the everyday life. The unique and superb performance of rubber tires has enabled automobiles to be accepted by ordinary people not necessarily by the professionals, which in turn necessitates some serious warnings from the SD viewpoints.

Presently, the competition between merits and demerits in the automobile-centered society is severe and somewhat complicated. On a few issues, exactly scientific or at least politically neutral conclusion is too difficult to reach. How we or the earth can accommodate all the automobiles whose number is rapidly increasing in almost all the corners in the world (not only in developed countries but also in most of developing countries) is a big problem to be solved soon for the survival of us on the earth. Inside the automobile industry, this problem has of course been discussed, and many technical developments have been achieved. Much different from railroads, cars are of course driven in various environments including residential areas. This advantage of automobile is due to its excellent performance (by using pneumatic rubber tires), and it is the reason of particular convenience for many people, including aged people, babies, and disabled. However, this versatility and familiarity of the cars in the everyday life has necessarily resulted in two negative influences: One is the scattering of carbon dioxide and a few other gases, and the second is constantly increasing traffic accidents, which is becoming more and more with the increasing number of automobiles in spite of strong campaigning to reduce traffic accidents on the roads. The rational solutions have still to be presented.

Using oil as a fuel of internal engine had been a big technical advantage of automobiles, but the polluted area might be often delocalized by the exhausted gases. At the same time, the emission of carbon dioxide from automobiles is one of the reasons of the global warming effect. Improvement in oil mileage has been constantly achieved by the continuous efforts done by the car makers, still being distant from zero emission. Change to electric vehicle (EV) is a reasonable choice, and industrial mass productions of EV are just beginning in a few countries.

However, still more than ten years may be necessary for the ubiquitous presence of EVs. Establishing electrical outlets for charging the secondary battery of EV has been agonizingly slow. Here again the delayed infrastructure or logistics is a big problem. As a compromise, so-called hybrid cars were developed, and now their share is increasing. Historically speaking, they turn out not to be a pinch hitter, but is occupying a definite share in the car market; that is, the hybrid cars are playing a role of good precursor of EVs now [26]. However, these technical improvements of the performance and new developments of various cars by the car industry seem not enough to overtake or overcome the big problems of ubiquitous traffic accidents as well as the global warming problems, definitely not possible in a couple of decade [26, 30, 37]. Now, we seemingly have to place a question, "Isn't it too late?"

6.3 Rubber Science and Rubber Technics of the Next Generation?

During the last few decades, two recognitions are popular on how to technically characterize the present period: One is Information Age and the other Transportation Age. The former might have enjoyed a majority support in general. For example, it is reported in an issue of TIME [37], that the cross-border exchange of information is increasing yearly (the bandwidth is increasing from gigabits to terabits in a recent few years), and in the article a financial expert has mentioned the increase of digital trades as the answer to question "Is there any good news?" However, one of the coauthors of this book has discussed this issue [26], to conclude that the transportation or logistics may be a predominant factor, presently and probably until the end of this century.

In terms of the increasing rate, the rate of automobile increase is smaller than that of the bandwidth of information movement. It is a popular belief that there would be no needs for meeting at a place by taking a trouble of coming and going of the attending people, if all the relevant people have their advanced mobile phones. However, this belief overlooks an important fact. Yes, there are many simple issues which may be dissolved by having chats on their mobiles without debate or discussion, but all are not so simple, of course.

We have much more issues too complicated to be solved by simply chatting on the phones. For concluding how to manage lots of problems in our life, in our community or in our office, definitely calling is not sufficient. A chat on the phone can only be the first step to deliver the problem itself. In our life or in our society, to give a good answer to a wrong or unimportant question is of no use. To give an answer, is much more critical to a good question, how dull or lackluster it may look to be. Therefore, even for establishing a problem itself, we need to meet together for lively discussions, when the issue is critically complex. The fundamental base of this conclusion may be a fact that we are humane. In order to overcome this factor, information technology of both hardware and software has to achieve much more advance to make it more humane or at least much more friendly to people, and such a progress might be extended well into the twenty-first century.

On the other hand, the transportation systems already have history of over 200 years, and they are maturing in terms of hardware. From the logistics and SD viewpoints, we still have to pay much more effort on rationally systematizing them, including betterment of computer software. As mentioned, automobiles are now presenting more number of urgent social and technical problems than the other transportation means, even though the traffic system is now at a mature stage. Related to this issue, two specific problems of automobiles in the transportation systems were already discussed [26]. That is, one problem is much specific to Asia where historically the horse carriages did not become much popular compared with Europe. There, the road systems were too poor to accommodate automobiles to result in the 24-h traffic jam in lots of metropolitan areas in Asia.

The second is introduced in Sect. 6.2, but to be extended here a little. For automobiles, being driven through people's living areas is common, except on the motorway. This feature induces striking difference from trains on the rail, ships in the sea, and flying aircrafts. Pneumatic rubber tires are a revolutionary invention that has enabled "horseless carriages" to be driven by ordinary people even through the lanes in the crowded residential area (see Sect. 2.4.2 and Chap. 5). However, the differences of weight and speed of automobiles from those of pedestrians (people) are too big to overcome the possibility of traffic accidents, in spite of the unique and excellent performance of rubber tires [35–40]. The driving of the automobiles is not necessarily by a professional driver is also tricky and problematic, from the urgent and critical object of reducing the terrible traffic accidents. There appeared lots of psychological studies focusing on the drivers, which may be on this consciousness [39]. Modifications of the license system for driving on this line may be a serious and urgent issue all over the world.

Recently, automatic driving systems are fashionable, based on the progress of artificial intelligence (AI). The AI research had its origin in the studies by C.E. Shanon (1916–2001) who had initiated a new scientific arena, information technology [26, 41]. So, AI may have its history not short, and the recent activity has its own solid base. The application of it to automobile driving techniques, however, may possibly be just at beginning, when the complicated dynamic situations encircling the driven cars on the town roads are considered with caution. For a moment, automobiles guided by the AI system may be safely used only on the motorways, or in the country-side areas where pedestrians are few. So far, the technics on automobile driving have been focused on the convenience and safety of drivers, not much on the safety of pedestrians. The future technics has to be centered much more on the safety of outside people, mainly pedestrians, especially a child, disabled, aged people, people carrying a baby, etc., are to be cared. The science and technics of pneumatic rubber tires are now matured, but what tire technics can do on this direction may not be high enough to give an excellent solution. Still, further contribution is expected from the tire side to the safety of car driving (see Sect. 5.5), which would be a critical issue associated with the transportation society from the SD standpoint.

Previous mention (see the last paragraph in Sect. 6.2) of the exhaust from the internal engine of automobiles is further discussed here. Reduction of the exhaust from the ubiquitous cars is one of the central issues for controlling the global warming, together with the local air pollution. Improvement in the exhaust reduction has a limit if stick to the internal engines using oil, and now hybrid cars and EV are highlighted. The former was marketed in the middle of 1990s, and the share is increasing, particularly in the developed countries, and the trend is recently followed by EV [26]. Here, EV is including PEV (plug-in EV), that is, the recharging of the secondary battery on the car is possible from the electrical sources at home. The development of decent secondary batteries is vital for EV, and the lithium-ion batteries are improved to show the promising performance [42, 43]. Establishing infrastructure for EV, particularly electrical charging station, is the most urgent issue to be solved. Marketing of PEV is expected to give a little time allowance before establishing sufficient numbers of the station. Anyway, the pneumatic rubber tires do not present any vital problems to this trend. FCVs (fuel-cell vehicles) are supposedly assumed to be one of the possible next generation cars after EV, and no vital problems are considered on the rubber tires even for FCVs.

If anything, the stable supply of natural rubber (NR), the most fundamental raw material of tires, may be enough or not in a near future is a concern among rubber specialists [44–47]. On the alternatives of *Hevea* tree for the NR source, see Sect. 3.4.2. At the same time, a worry on the damage to the tropical rain forests by the expansion of *Hevea*-tree plantations has to be taken into account [44, 45, 47]. Novel rational way of increased production of NR is to be further developed from the SD standpoint [45].

From the viewpoint of SD, a lot of problems about automobiles are still to be considered, but at least the rubber as a material of automobile tire is concerned, it is notable that the present level of rubber tire technology would manage to react appropriately for a moment. Rubber science and rubber technics are now destined to tackle some future problems as well as the urgent facing problems. Consequently, we will have to elucidate the future issues to be studied, and three major assignments are tentatively proposed here:

Firstly, half a century has elapsed after the establishment of the paradigm of rubber vulcanization in 1970s (see Sect. 2.3 and Refs. [48–50]), thus new sulfur vulcanization paradigm might be considered in a near future (see Sect. 4.3).

Secondly, technology and technics of pneumatic rubber tires are already at the matured stage (see Chap. 5), and a breakthrough of engineering design of tires is hoped in mechanical engineering arena.

Thirdly, Resource of NR is too much dependent on *Hevea brasiliensis* to secure a stable supply of NR products including tires (see Sect. 3.4.2 and Refs. [44–47]). How to dissolve the biodiversity and biosecurity issue is subject to discussions among rubber-related people.

The above three issues suggest that rubber science and rubber technics are yet to face the emergency in order to secure the SD in the Transportation Society. In other words, rubber engineers may have a little more time in finding the solutions of these issues. It may be fortunate also for rubber scientists to have some time, which, however, should mean giving fundamentally right solutions is mandatory, i.e., not a temporary answer but an exact response has to be given: Fortunately, on the first issue, the first arrow has been shot [48–50]. If this arrow turns out to be a breakthrough, the further development would possibly lead to the new vulcanization paradigm in a due course (see Fig. 4.17). This process to the new paradigm must be hard and complicated. There are to be lots of ups and downs, and it would be a challenge for rubber scientists and engineers. We are looking forward to the paradigm change in vulcanization arena hopefully well before the end of this century.

On the second issue, only a little discussion has been given in Sect. 5.5.2. Some original ideas that are not at the extension of the past technics are to be proposed [34], when considering the maturity of the present engineering research. A breakthrough is in need. The third one is still not much uncovered, even though a stimulating article appeared in *Chemical & Engineering News* from American Chemical Society [46], and a review is disclosed [44, 45]. In the USA and Europe, the issue is more popular than in Asia, where most *Hevea* trees are cultured and are currently providing major amount of NR. Under these situations, Japan situated in Asia has to act immediately for the future stable supply of NR. A recent report may pioneer the way to in vitro synthesis of natural rubber [51], which is expected to be one of the important contributions from Japan representing a different approach from those in the USA and Europe.

This book is a trial of modern approach to rubber science, not necessarily giving a solution to technical troubles during the manufacturing, processing, and utilization of rubbers. However, without a modernization of rubber science in this century when the transition of the transportation society to the information society is on the way, many problems might remain to be unsolved or the solutions may be delayed. History suggests that both science and technics themselves are to be continued. However, after the matured or stabilized period, technological innovation is a must for the survival of human beings. We, all the authors sincerely hope this book is followed by more trials toward the modern approach of rubber science, from a different standpoint from us. Only combined with similar trials by other groups, we might hope for the establishment of modern rubber science well before the next century.

Remark 5 Nobel Prize: The Good and Bad Sides

Every year early in October, science is focused among mass communication media in many countries. The reason why? Of course, the announcement of the Nobel Prize winners in physics, chemistry, and physiology or medicine is going to be disclosed. In each winner's country, the news appears at the top, since it should be much delightful to have the winner, which informs the world of the scientific excellence. Most people would be bewildered if questioned from a scientist, "Has the winner contributed so much to the progress of science?" That is none of their concerns, and the scientist should have known it.

   In cooperation with several researchers, American and European societies of physics jointly summarized the physical research activities in twentieth century. The publication listed 63 physicists, who have made most valuable contribution to the progress in physics [52]. Among 63 physicists, there find two Japanese, H. Yukawa (a Nobel Prize winner) and R. Kubo (not a winner), and 37 are the Nobel Prize winners (physics, 29 and chemistry, 8). Accordingly, the ratio of the Nobel Prize winners is 58.7% (=37/63). Is this ratio high? Or, low? Many may feel this ratio is unexpectedly low, especially who have assumed the highest honor to the Nobel Prizes in the scientific fields, perhaps.

   However, more important is the other ratio. The number of winners of Nobel Prize in physics was 140 (between 1900 and 1990, one received twice is counted as two). Among 140 Nobel laureates, only 29 (or 37 if the winners of chemistry are included) are chosen among the maximal contributors to the progress in physics. Therefore, 111 (=140 − 29, 79.3%) or 103 (=140 − 37, 73.6%) is the number of physicists (and the ratio) who were not chosen to be the maximal contributors among 140 laureates. This calculation suggests that when tracing the progress of physics, to list the achievements by the Nobel Prize winners and to explain them in details is absolutely insufficient, and it cannot escape the criticism of being biased. In other words, there are many (or too many?) research results of value to be studies carefully outside of the coverage of the Nobel Prizes. This can be the case not only in physics but also in the other areas of science. We probably have to be much more careful in evaluating the Nobel Prizes in scientific fields.

   As explained in Sect. 2.4.3, the first Nobel laureate from polymer chemistry field, Prof. Staudinger, researched chemical reactions of rubber extensively, in order to verify his macromolecular theory. The second winner, Prof. Ziegler, pioneered the way to new synthetic rubbers (EPM, EPDM, and a few other polyolefin rubbers). The third, Prof. Flory, is one of the founders of polymer science by his contribution to rubber elasticity theory. In these cases, roles of rubber science are essential. Before the Transportation Age turns into the Information Age, the winner from rubber science field itself is hoped to appear through the contribution to rubber science and sustainable development, both of which are surely of utmost importance for our future on the earth.

References

1. D.H. Meadows, D.L. Meadows, J. Randers, W.W. Behrens III, *The Limits to Growth: A Report for THE CLUB OF ROMES Project on the Predicament of Mankind* (Universe Books, New York, 1972)
2. C.F. von Weizsäsker, *Die Geschichte Der Natur* (Hirzel AG, Stuttgart, 1958)
3. A. Schmidt, *Der Begriff Der Natur in Der Lehre von Marx* (Europäische Verlagsanstalt, Frankfurt am Main, 1962)
4. C.H. Langmuir, W. Broecker, *How to Build a Habitable Planet: The Story of Earth from the Big Bang to Humankind* (Princeton University Press, Princeton, NJ, 2012)

5. F. Dannemann, *Die Naturwissenschaften in ihrer Entwicklung und Zusammenhange* (Die Zweite Auflage in Vier Bänden, Leipzig, 1920–1923) [Complete translation into Japanese is available in Japan]

6. P. Bellwood, *First Farmers: The Origin of Agricultural Societies* (Blackwell Publishing, Oxford, 2005)

7. T.K. Derry, T.I. Williams, *A Short History of Technology: From the Earliest Times to A.D. 1900* (Oxford University Press, Oxford, 1960)

8. L. Beck, *Die Geschichite des Eisens in Technischer und Kulturgeschichtlicher Beziehung, I-V* (Friedrich Vieweg und Sohn, Braunschweig, 1884–1903) [Complete translation into Japanese is available in Japan]

9. S. Mason, *A History of the Sciences: Main Currents of Scientific Thought* (Abelard-Schuman, London, 1953)

10. C. Ponting, *A Green History of the World* (Penguin Books, London, 1992)

11. J. Mokyr, *The Lever of Riches: Technological Creativity and Economic Progress* (Oxford University Press, New York, 1990)

12. H.J. Störig, *Kleine Weltgeschichte der Wissenschaft* (Kohlhammer, Stuttgart, 1954)

13. H. Buttefield, *The Origin of Modern Science, 1300–1800* (Bell, London, 1949)

14. D.C. Linberg, R.S. Westman (eds.), *Reappraisals of the Scientific Revolution* (Cambridge University Press, Cambridge, 1990)

15. F. Borkenau, *Der Übergang vom Feudalen zum Bürgerlichen Weltbild: Studien zur Geschichte der Philosophie der Manufacturperiode* (Felix Alcan, Paris, 1934)

16. L. Mumford, *The Transformation of Man* (Harper & Row, New York, 1956)

17. C. Singer, *A Short History of Scientific Ideas to 1900* (Clarendon Press, Oxford, 1959)

18. T.S. Ashton, *The Industrial Revolution, 1760–1830* (Greenwood Press, Westport, 1986)

19. W.H.G. Armytage, *A Social History of Engineering* (Faber and Faber, London, 1961)

20. E. Hobsbawn, *The Age of Revolution 1789–1848* (Vintage Books, New York, 1962)

21. D.R. Headrick, *The Tool of Empire: Technology and European Imperialism in the Nineteenth Century* (Oxford University Press, New York, 1981)

22. L. Turner, W.C. Clark, R.W. Kates (eds.), *The Earth as Transformed by Human Action: Global and Regional Changes in the Biosphere over the Past 300 years* (Cambridge University Press, Cambridge, 1990)

23. D. Arnold, *The Problem of Nature: Environment, Culture and European Expansion* (Blackwell Publishers, Oxford, 1996)

24. M.K. Matossian, *Shaping World History: Breakthroughs in Ecology, Technology, Science, and Politics* (M. E. Shape, Armonk, 1997)

25. A. Weissman, *The World without Us* (Thomas Dunne Books, New York, 2007)

26. S. Kohjiya, *Natural Rubber: From the Odyssey of the Hevea Tree to the Age of Transportation* (Smithers Rapra, Shrewsbury, 2015)

27. D.P. Miller, *James Watt, Chemist: Understanding the Origins of the Steam Age* (Pickering & Chatto, London, 2009)

28. D. Brinkley, *Wheels for the World: Henry Ford, His Company, and a Century of Progress, 1903–2003* (Viking, New York, 2003)

29. B.S. Katz, *Infrastructure*, in *Encyclopedia of Economics* (McGraw-Hill, New York, 1982), pp. 523 [Katz noted that this word was used in NATO's documents in early 1950s]

30. T. Jouenne, Sustainable logistics as a part of modern economies, in *Sustainable Solutions for Modern Economics*, ed. by R. Hoefer (RSC Publishing, Cambridge, 2009), Chap. 4

31. L. Birmingham, D. McNeill, *Strong in the Rain: Surviving Japan's Earthquake Tsunami, and Fukushima Nuclear Disaster* (Palgrave Macmillan, New York, 2012)

32. World Commission on Environment and Development, *Our Common Future* (Oxford University Press, Oxford, 1987)

33. E.B. Barbier, *Natural Resources and Economic Development* (Cambridge University Press, Cambridge, 2005)

34. J.W. Tester, E.M. Drake, M.W. Golay, M.J. Driscoll, W.A. Peters, *Sustainable Energy: Choosing among Options* (MIT Press, Cambridge, MA, 2005)

35. E. Millstone, T. Lang (eds.), *The Atlas of Food: Who Eats, Where, and Why* (Myriad Edition, Brighton, 2003)
36. E. Dinjus, U. Arnold, N. Dahmen, R. Hoefer, W. Wach, Green fuels-sustainable solutions for transportation (Chap. 8), in *Sustainable Solutions for Modern Economics*, ed. by R. Hoefer (RSC Publishing, Cambridge, 2009)
37. J. Grant, *TIME*, April 25, p. 37 (2016)
38. H. Uzawa, *Jidoushano Shakaitekihiyou (Social Costs of Automobiles)* (Iwanami Shoten, Tokyo, 1974) [In Japanese]
39. G. Underwood (ed.), *Traffic and Transport Psychology: Theory and Application* (Elsevier, Amsterdam, 2005)
40. D. Shinar, *Traffic Safety and Human Behavior* (Emerald Group Publishing, Bingley, 2007)
41. L. Brillouin, *Science and Information Theory* (Academic Press, New York, 1962)
42. S. Kohjiya, Y. Ikeda, Recent Research Developments in Electrochemistry **4**, 99 (2001)
43. T. Minami, M. Tatsumisago, M. Iwakura, S. Kohjiya, I. Tanaka (eds.), *Solid State Ionics for Batteries* (Springer, Tokyo, 2005)
44. S. Kohjiya, Nippon Gomu Kyokaishi, **88**, 93 (2015) [In Japanese]
45. Y. Ikeda, A. Tohsan, S. Kohjiya, in *Sustainable Development: Processes, Challenges and Prospects*, ed. by D. Reyes (Nova Science Publishers, New York, 2015), Chap. 3
46. A.H. Tullo, Chem. Eng. News, April 20, 18 (2015)
47. C. Roberts, TIME, July 25, 21 (2016)
48. Y. Ikeda, Y. Yasuda, T. Ohashi, H. Yokohama, S. Minoda, H. Kobayashi, T. Honma, Macromolecules **48**, 462 (2015)
49. Anon, *New Insight into How Rubber Is Made Could Improve Tires, Reduce Air Pollution, PresPac*, Feb. 11. (A weekly press release journal from American Chemical Society, Washington, D.C., 2016)
50. Y. Ikeda, H. Kobayashi, S. Kohjiya, Kagaku (Chemistry) **70**, 19 (2015) [In Japanese]
51. S. Yamashita, H. Yamaguchi, T. Waki, Y. Aoki, M. Mizuno, F. Yanbe, T. Ishii, A. Funaki, Y. Tozawa, Y. Yamagi-Inoue, K. Fushihara, T. Nakayama, S. Takahashi. eLife **5**, e19022 (2016)
52. L.M. Brown, A. Pais, B. Pippard (eds.), *Twentieth Century Physics, published in 3 volumes* (Institute of Physics Publishing and American Institute of Physics Press, New York, 1995)

Appendix

The table of commercially available rubbers (quality evaluation: A, superior; B, good; C, fine; D, inferior)

Rubber		Natural rubber	Isoprene rubber	Styrene-butadiene rubber	Butadiene rubber
Abbreviation (ASTM)		NR	IR	SBR	BR
Raw rubber	Specific gravity	0.91~0.93	0.92~0.93	0.92~0.97	0.91~0.94
	Mooney viscosity ML_{1+4} (100 °C)	45~150	55~90	30~70	35~55
	Solubility parameter (SP)	7.9~8.4	7.9~8.4	8.1~8.7	8.1~8.6
Physical performances	Tensile strength (MPa)	3~35	3~30	2.5~30	2.5~20
	Elongation (%)	100~1000	100~1000	100~800	100~800
	Hardness (JIS:A)	10~100	20~100	30~100	30~100
	Resilience	A	A	B	A
	Tear strength	A	B	C	B
	Compression set	A	A	B	B
	Flex resistance	A	A	B	C
	Abrasion resistance	A	A	A	A
	Aging resistance	B	B	B	B
	Ozone resistance	D	D	D	D
	Light resistance	B	B	B	D
	Flame resistance	D	D	D	D

(continued)

(continued)

Rubber		Natural rubber	Isoprene rubber	Styrene-butadiene rubber	Butadiene rubber
Abbreviation (ASTM)		NR	IR	SBR	BR
	Gas retention	B	B	C	B
	Radiation resistance	C	C	B	D
	Electronic insulation (Ω cm)	$10^{10} \sim 10^{14}$	$10^{10} \sim 10^{15}$	$10^{10} \sim 10^{15}$	$10^{14} \sim 10^{15}$
	Upper service temperature (°C)	120	120	120	120
	Lowest service temperature (°C)	$-50 \sim -70$	$-50 \sim -70$	$-40 \sim -65$	-70
Solvent resistance	Gasoline, light oil	D	D	D	D
	Benzene, toluene	D	D	D	D
	Alcohol	A	A	A	A
	Ethyl acetate	C	C	C	C
Acid and alkali resistances	Organic acid	D	D	D	D
	High concentration of mineral acid	C	C	C	C
	High concentration of alkali	B	B	B	B
Principal end use		Large tire, Foot wear, Hose, band, Various rubber products	Tire, Foot wear, Hose, band, Various rubber products	Tire, Foot wear, Rubber cloth, Floor tile, Battery case, Various rubber products	Antivibration rubber, Foot wear, Tire, Various rubber products
Chemical structure	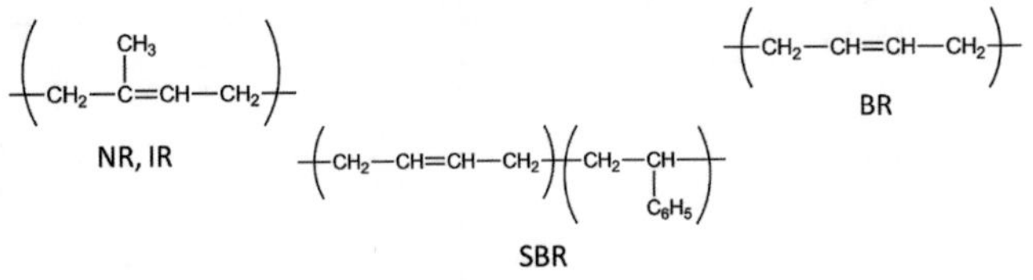				

$$\left(CH_2 - \underset{\underset{CH_3}{|}}{C} = CH - CH_2 \right)$$

NR, IR

$$\left(CH_2 - CH = CH - CH_2 \right)\left(CH_2 - \underset{\underset{C_6H_5}{|}}{CH} \right)$$

SBR

$$\left(CH_2 - CH = CH - CH_2 \right)$$

BR

Rubber		Polychloroprene	Butyl rubber	Ethylene–propylene rubber	Ethylene–vinyl acetate copolymer	Chlorosulfonated polyethylene
Abbreviation (ASTM)		CR	IIR	EPM, EPDM	EAM	CSM
Raw rubber	Specific gravity	$1.15 \sim 1.25$	$0.91 \sim 0.93$	$0.86 \sim 0.87$	$0.98 \sim 0.99$	$1.11 \sim 1.18$
	Mooney viscosity ML_{1+4} (100 °C)	$45 \sim 120$	$45 \sim 75$	$40 \sim 100$	$20 \sim 30$	$30 \sim 115$
	Solubility parameter (SP)	$8.2 \sim 9.4$	$7.7 \sim 8.1$	$7.9 \sim 8.0$	$7.8 \sim 10.6$	$8.1 \sim 10.6$
Physical performances	Tensile strength (MPa)	$5 \sim 25$	$5 \sim 20$	$5 \sim 20$	$7 \sim 20$	$7 \sim 20$
	Elongation (%)	$100 \sim 1000$	$100 \sim 800$	$100 \sim 800$	$100 \sim 600$	$100 \sim 500$
	Hardness (JIS:A)	$10 \sim 90$	$20 \sim 90$	$30 \sim 90$	$50 \sim 90$	$50 \sim 90$
	Resilience	B	D	B	B	C
	Tear strength	B	B	C	B	B
	Compression set	B	C	B	D	C
	Flex resistance	B	A	B	B	B
	Abrasion resistance	A	B	B	A	A
	Aging resistance	A	A	A	A	A
	Ozone resistance	A	A	A	A	A
	Light resistance	A	A	A	A	A
	Flame resistance	B	D	D	B	B
	Gas retention	B	A	B	B	A
	Radiation resistance	C	D	C	C	C
	Electronic insulation (Ω cm)	$10^{10} \sim 10^{12}$	$10^{15} \sim 10^{18}$	$10^{12} \sim 10^{16}$	$10^{12} \sim 10^{14}$	$10^{12} \sim 10^{14}$
	Upper service temperature (°C)	130	150	150	200	150
	Lowest service temperature (°C)	$-30 \sim -40$	$-30 \sim -60$	$-40 \sim -60$	$-20 \sim -30$	$-20 \sim -60$

(continued)

(continued)

Solvent resistance	Gasoline, light oil	B	D	D	B	C
	Benzene, toluene	D	C	C	D	D
	Alcohol	A	A	A	A	A
	Ethyl acetate	D	A	A	C	D
Acid and alkali resistances	Organic acid	D	C	D	D	C
	High concentration of mineral acid	B	A	B	B	A
	High concentration of alkali	A	A	A	A	A
Principal end use		Electric wire, Conveyor belt, Antivibration rubber, Rubber for window frame, Adhesive, Paint	Tire inner tube, Electric wire, Steam hose, Heat-resistant conveyer belt	Electric wire, Rubber for window frame, Hose, Conveyor belt	Heat-resistant gasket	Lining, Weather and corrosion resistant paint, Rubber cloth for outdoor use
Chemical structure						

Rubber		Chlorinated polyethylene	Epichlorohydrin rubber	Nitrile butadiene rubber
Abbreviation (ASTM)		CM	CO, ECO	NBR
Raw rubber	Specific gravity	$1.10 \sim 1.20$	$1.27 \sim 1.36$	$1.00 \sim 1.20$
	Mooney viscosity ML_{1+4} (100 °C)	$68 \sim 76$	$35 \sim 120$	$30 \sim 100$
	Solubility parameter (SP)	$8.6 \sim 8.8$	$9.6 \sim 9.8$	$8.7 \sim 10.5$
Physical performances	Tensile strength (MPa)	$7 \sim 20$	$7 \sim 15$	$5 \sim 25$
	Elongation (%)	$100 \sim 600$	$100 \sim 500$	$100 \sim 800$
	Hardness (JIS:A)	$50 \sim 85$	$20 \sim 90$	$20 \sim 100$
	Resilience	C	B	B
	Tear strength	B	B	C
	Compression set	C	B	B
	Flex resistance	B	B	B
	Abrasion resistance	A	B	B
	Aging resistance	A	A	B
	Ozone resistance	A	A	D
	Light resistance	A	A	C
	Flame resistance	B	B	D
	Gas retention	B	A	B
	Radiation resistance	–	–	C
	Electronic insulation (Ω cm)	$10^{11} \sim 10^{12}$	$10^{9} \sim 10^{10}$	$10^{10} \sim 10^{11}$
	Upper service temperature (°C)	160	170	130
	Lowest service temperature (°C)	$-10 \sim -30$	$-20 \sim -40$	$-40 \sim -50$
Solvent resistance	Gasoline, light oil	B	A	A
	Benzene, toluene	D	D	C
	Alcohol	A	A	A
	Ethyl acetate	D	D	D
Acid and alkali resistances	Organic acid	B	C	C
	High concentration of mineral acid	A	B	B
	High concentration of alkali	A	B	B

(continued)

(continued)

Principal end use	Chemical-resistant hose, Roll, Lining	Inner tube, Oil seal, Oil-resistant hose	Gasket, Oil-resistant rubber products, Oil seal, Printer roll
Chemical structure	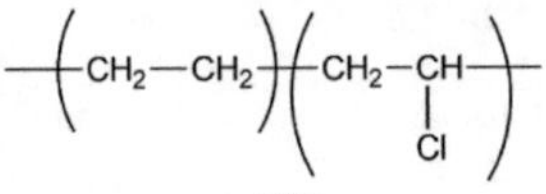 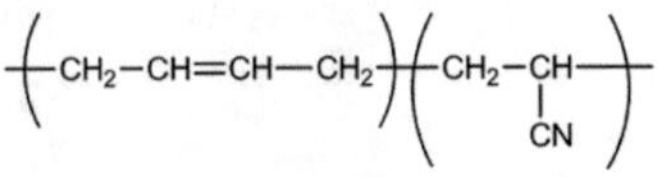 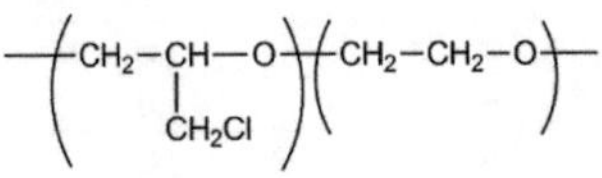 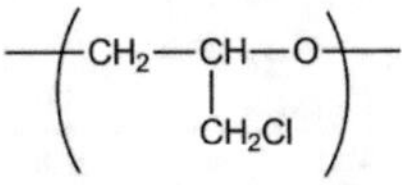		

$$-\left(CH_2-CH_2\right)\left(\begin{array}{c}CH_2-CH-\\ |\\ Cl\end{array}\right)-$$

CM

$$-\left(CH_2-CH=CH-CH_2\right)\left(\begin{array}{c}CH_2-CH-\\ |\\ CN\end{array}\right)-$$

NBR

$$-\left(\begin{array}{c}CH_2-CH-O\\ |\\ CH_2Cl\end{array}\right)\left(CH_2-CH_2-O\right)-$$

ECO

$$-\left(\begin{array}{c}CH_2-CH-O\\ |\\ CH_2Cl\end{array}\right)-$$

CO

Rubber		Acrylic rubber	Urethane rubber	Polysulfide rubber	Silicone rubber	Fluorine rubber
Abbreviation (ASTM)		ACM	U	T	Q	FKM
Raw rubber	Specific gravity	$1.09 \sim 1.10$	$1.00 \sim 1.30$	$1.34 \sim 1.41$	$0.95 \sim 0.98$	$1.82 \sim 1.85$
	Mooney viscosity ML_{1+4} (100 °C)	$45 \sim 60$	$25 \sim 60$ (or liquid state)	$25 \sim 50$ (or liquid state)	Liquid state	$35 \sim 160$
	Solubility parameter (SP)	9.4	10.0	$9.0 \sim 9.4$	$7.3 \sim 7.6$	8.6
Physical performances	Tensile strength (MPa)	$7 \sim 15$	$20 \sim 45$	$3 \sim 15$	$3 \sim 15$	$7 \sim 20$
	Elongation (%)	$100 \sim 600$	$300 \sim 800$	$100 \sim 700$	$50 \sim 500$	$100 \sim 500$
	Hardness (JIS:A)	$40 \sim 90$	$60 \sim 100$	$30 \sim 90$	$30 \sim 90$	$50 \sim 90$
	Resilience	C	A	C	B	C
	Tear strength	C	A	C	D	C
	Compression set	B	A	D	A	B
	Flex resistance	B	A	D	C	D
	Abrasion resistance	B	A	C	C	A
	Aging resistance	A	B	A	A	A
	Ozone resistance	A	A	A	A	A
	Light resistance	A	A	A	A	A
	Flame resistance	D	C	D	C	A
	Gas retention	B	B	A	C	A
	Radiation resistance	C	B	C	B	C
	Electronic insulation (Ω cm)	$10^8 \sim 10^{10}$	$10^9 \sim 10^{12}$	$10^{13} \sim 10^{15}$	$10^{11} \sim 10^{16}$	$10^{10} \sim 10^{14}$
	Upper service temperature (°C)	180	80	80	280	300
	Lowest service temperature (°C)	$-10 \sim -30$	$-30 \sim -60$	$-10 \sim -40$	$-70 \sim -120$	$-10 \sim -50$

(continued)

(continued)

Solvent resistance	Gasoline, light oil	A	A	A	C	A
	Benzene, toluene	D	D	B	C	A
	Alcohol	D	C	A	A	A
	Ethyl acetate	D	C	B	C	D
Acid and alkali resistances	Organic acid	D	D	D	B	D
	High concentration of mineral acid	C	D	D	C	A
	High concentration of alkali	C	D	C	A	D
Principal end use		Heat- and oil-resistant rubber products, Oil seal	Rubber seal for high pressure, Solid tire, Track tie pad	High oil resistant Hose, rubber roll, Rubber seal	Insulating rubber products for electronics, Rubber products for very high- and low-temperature uses	Rubber seal with heat, Oil, and Chemical resistances

Chemical structure

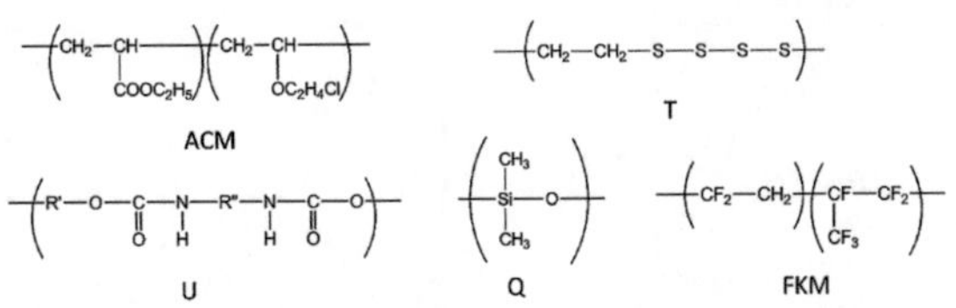

$\left(\!CH_2\!-\!CH\!\right)_{}\!\left(\!CH_2\!-\!CH\!\right)_{}$
$\qquad COOC_2H_5 \qquad OC_2H_4Cl$

ACM

$\left(\!R'\!-\!O\!-\!\underset{O}{\overset{}{C}}\!-\!\underset{H}{\overset{}{N}}\!-\!R''\!-\!N\!\underset{H}{\overset{}{}}\!-\!\underset{O}{\overset{}{C}}\!-\!O\!\right)$

U

$\left(\!CH_2\!-\!CH_2\!-\!S\!-\!S\!-\!S\!-\!S\!\right)$

T

$\left(\!\underset{CH_3}{\overset{CH_3}{Si}}\!-\!O\!\right)$

Q

$\left(\!CF_2\!-\!CH_2\!\right)\!\left(\!\underset{CF_3}{\overset{}{CF}}\!-\!CF_2\!\right)$

FKM

Index

Printed in the United States
By Bookmasters